The In-Memory Revolution

How SAP HANA Enables Business of the Future

HASSO PLATTNER & BERND LEUKERT

THE IN-MEMORY REVOLUTION

HOW SAP HANA ENABLES BUSINESS OF THE FUTURE

 Springer

Hasso Plattner
Hasso Plattner Institute
Potsdam
Germany

Bernd Leukert
SAP SE
Walldorf
Germany

ISBN 978-3-319-16672-8
DOI 10.1007/978-3-319

Library of Congress Control Number: 2015935727
Springer Cham Heidelberg New York Dordrecht London
© Springer International Publishing Switzerland 2015

Printed on acid-free paper

Springer International Publishing AG Switzerland is part of Springer Science+Business Media (www.springer.com)

Disclaimer

In several sections of this book, we use generalizations for the sake of understandability or future prognoses. Such parts only outline possibilities and plans for potential future improvements, with no guarantee that they will indeed come to pass at any point in the future. This book is not official SAP communication material. Any decisions you make should be based on official SAP communication material. All statements concerning the future are subject to various risks and uncertainties that could cause actual results to differ materially from expectations. Readers are cautioned not to place undue reliance on these statements, which speak only as of their dates, and they should not be relied upon in making purchasing decisions. This book is meant to give you an impression of how SAP HANA and S/4HANA could improve your business. Some of the applications or features mentioned throughout the book require separate licenses.

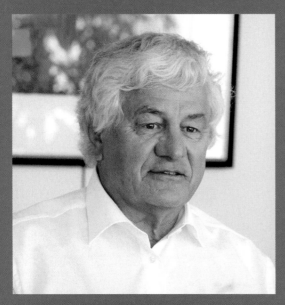

Credit: Berthold Steinhilber / laif

Prof. Dr. h.c. Hasso Plattner

Hasso Plattner Institute (HPI), University of Potsdam, Germany

Prof. Dr. h.c. Hasso Plattner is the chair of the Enterprise Platform and Integration Concepts research group at the HPI, which focuses mainly on in-memory data management for enterprise applications and human-centered software design and engineering. He is the author or co-author of over 100 scientific publications and has served as a keynote speaker in conferences such as ACM SIGMOD and VLDB. He currently also serves as a visiting professor at the Stanford University Institute of Design. Prof. Plattner received his diploma in communications engineering from the University of Karlsruhe. In recent years, he has been focusing on teaching and research in the field of business computing and software engineering at large. In 1998, he founded the HPI in Potsdam, Germany. At the HPI, approximately 500 students are currently pursuing their Bachelor and Master degrees in IT Systems Engineering with the help of roughly 50 professors and lecturers. The HPI currently has over 100 PhD candidates.

Credit: Ingo Cordes / SAP SE

Bernd Leukert

Member of the Executive Board, SAP SE

Bernd Leukert is a member of the Executive Board and the Global Managing Board of SAP. He is responsible for the development and delivery of all products across SAP's portfolio, including applications, analytics, cloud, mobile, as well as database and technology. Leukert joined SAP in 1994 and has held various management positions in application development, technology development, software engineering, and process governance. Leukert's efforts have contributed considerably to SAP's product portfolio, including SAP R/3, Supply Chain Management (SCM), SAP Business Suite, SAP Business One, and the SAP Business Suite 4 SAP HANA (S/4HANA).

Contact

Website: http://www.in-memory-revolution.com
Email address: in-memory-revolution-book@sap.com

THE TEAM

Editors

Mohammed AbuJarour, Zbigniew Jerzak, Daniel Johannsen, Michael Perscheid, and Jürgen Müller

Co-Editors

Adelya Fatykhova, Pablo Guerrero, Josefine Harzmann, Martin Heinig, Lisa Nebel, Martin Przewloka, and Ralf Teusner

Designers

Amid Ayobi, Kirsten Grass, Elisabeth Kaminiczny, Nadine Krauß-Weiler, Lars Kreuzmann, John Pompa, and Maciej Szaniawski

Contributors

We would like to thank the following people for their contributions to the examples contained in this book:

Andrea Anderson, Jens Baumann, Pavan Bayyapu, Jens Bensing, Philipp Berger, Natalie Bernzen, Dominik Bertram, Jim Brooks, David Burdett, Christian Busch, Arturo Buzzalino, John Carter, Nalini Chandhi, Piyush Chandra, Fredrick Chew, Manny Cortez, Tobias Decker, Cindy Fähnrich, Daniel Faulk, Barbara Flügge, Steven Garcia, Boris Gelman, Hinnerk Gildhoff, Tim Grouisborn, Raja Gumienny, Gerhard Hafner, Thomas Heinzel, Patrick Hennig, Pit Humke, Abhishek Jaiswal, Jaehun Jeong, Thorsten Jopp, Stefan Kätker, Gerrit Simon Kazmaier, Priyanka Khaitan, Chung Rim Kim, Nicolas Liebau, Christoph Meinel, Holger Meinert, Jörg Michaelis, Stephan Müller, Bastian Nominacher, Franz Peter, Mario Ponce, Keertan Rai, Michael Rey, Alexander Rinke, Jan Schaffner, Karsten Schierholt, Dave Schikora, Daniela Schweizer, Heiko Steffen, Matthias Steinbrecher, James Tarver, Holger Thiel, Gregor Tielsch, Rohit Tripathi, Matthias Uflacker, Stefan Uhrig, Ramshankar Venkatasubramanian, Priya Vijayarajendran, Vaibhav Vohra, Lars Volker, Guido Wagner, Sebastian Walter, Wolfgang Weiss, Felix Wente, Tobias Wieschnowsky, Andreas Wildhagen, and Ruxandra Zamfir

Reviewers

We would like to thank the following people who provided valuable feedback on the content of this book:

Malte Appeltauer, Thomas Augsten, Upendra Barve, Deepak Bhojwani, Martin Boissier, Lars Butzmann, Sushovan Chattaraj, Robert Chu, Holger Dieckmann, Ofer Eisenberg, Morten Ernebjerg, Martin Faust, Valentin Flunkert, Florian Frey, Thomas Gawlitza, Lutz Gericke, Helge Gose, Henning Heitkötter, Peter Hoffmann, Matthias Kaup-Antons, Stefan Klauck, Thomas Kowark, Christian Krause, Nikolai Kulesza, Robert Kupler, Martin Lorenz, Stefanie Maute, Carsten Meyer, Pablo Paez, Oleksandr Panchenko, Andre Pansani, Enrique Garcia Perez, Keven Richly, Stefan Scheidl, Christian Schnetzer, David Schwalb, Christian Schwarz, Thomas Uhde, Martin Wezowski, Sebastian Wieczorek, Marc-Alexander Winter, and Sebastian Woinar

Acknowledgements

We would like to thank the following people who provided their support during the writing of this book:

Matthias von Blohn, Alexander Böhm, Michael Brown, Mathias Cellarius, Quentin Clark, Achim Clemens, David Dobrin, Michael Emerson, Markus Fath, Bernhard Fischer, Caroline Gotzler, Sundar Kamak, Mark von Kopp, Marcus Krug, Jens Krüger, Nicola Leske, Markus Noga, Franz Peter, Tanja Rueckert, Ann-Sofie Ruf, Karsten Schmidt, Henning Schmitz, Wieland Schreiner, Bertram Schulte, Jun Shi, Luisa Silva, Guruprasad Srinivasamurthy, Peter Weigt, Johannes Wöhler, Sam Yen, and Wenwen Zhou

Furthermore, we would like to acknowledge our co-innovation partners:

Celonis, ConAgra, Deutsche Telekom, Hamburg Port Authority, Hasso Plattner Institute, Home Shopping Europe 24, McLaren, National Center for Tumor Diseases in Heidelberg, and Siemens

Last but not least, we would like to thank all our SAP colleagues for making HANA and S/4HANA successful.

Contents

Part III   KEY BENEFITS OF HANA FOR ENTERPRISE APPLICATIONS   146

Foreword

by Prof. Clayton M. Christensen
Kim B. Clark Professor of Business Administration at the Harvard Business School

Sometimes life-changing events seem to come at times and from directions you didn't anticipate. One of these happened to me in 2014. I was minding my own business in my office at the Harvard Business School when the phone rang. It was Hasso Plattner, a founder of the massive European software company SAP. I had read about Plattner for sure – but, we had never met. Without asking whether I had the time, he simply started to talk. And why should I ask otherwise? For me, it was a chance of a lifetime – to listen to what one of the world's software geniuses wanted to tell me.

The arrival of Plattner's message was an amazing coincidence. I am a religious person, and I regularly think about whether God is pleased with my life. In one of these ponderings recently, I had an important insight: God does not need accountants in Heaven. Because we have finite minds, we need to aggregate data into bigger numbers to have a sense for what is going on around us. For example, I can't keep track of all of the specific invoices we have sent to our customers. So thank goodness, we have an accountant who can count up all these into a single number which we call "sales." We also receive additional numbers for individual bookings, costs, assets, liabilities, and so on. Without these numbers, or aggregates, data about our lives would be too complicated to understand. For better or worse, the need to aggregate often imparts to our lives a sense of hierarchy: people who preside over bigger numbers are more important in this world compared with people who preside over smaller numbers. Hence, the corporate ladder. I realized, however, that because God has an infinite mind, he doesn't need to aggregate above the level of individuals in order to have a perfect understanding of what is going on in the world. And this implies that when he measures my life, he will only discuss with me what I have done to help other people – because he doesn't aggregate above the level of the individual.

With these thoughts in my mind, I was stunned that after I explained this observation to Plattner, he said that he had seen the same pattern in ERP software. The architectural core of any data storage system is aggregates. Because we have finite minds, we need accountants and their computers to summarize the detailed pieces of data as aggregates. Periodically, we use algorithms to add, subtract, multiply, and divide these aggregates to convey a sense of what is going on in the enterprise.

Aggregates help executives to manage an enterprise. However, they impede executives' attempts to change the enterprise. How does this happen? I observe that there are four stages in the development of companies. The first stage is the company's first product that is comprised of components. In the second stage, the company creates groups of employees who have the responsibility to develop better components in the company's successful product or service. Hence, the groups in which employees work become a reflection of the architecture of the product. Third, as the company grows successfully, organization charts are developed to clarify what people and what groups are responsible for, and who they are accountable to. And fourth, people who are responsible for each box in the organization chart need to get data on what is going on in their box – and aggregates are defined to provide the executives the data they need.

As long as the aggregates are a faithful representation of an enterprise and its markets, an aggregate-based accounting system gives managers a useful but static view of reality. But the architecture of aggregates doesn't give us a dynamic view of our companies or our markets. The only way to get dynamics is to wait until the next static report is released. ERP systems cannot signal new trends in a company or its markets because the individual pieces of data which comprise data about the new direction are hidden in aggregates that were designed to reflect the old. Ferreting out the dispersed data which signals the future is very difficult, and is a key reason why established leaders are slow in innovation.

I have no financial stake in SAP's financial success. But from an academic perspective, SAP's new HANA system brings to management things that – even today – were unthinkable. The reason is that HANA does not materialize aggregates. It is as if the data are marbles, and are thrown on a perfectly flat floor. No piece of data is hidden under another piece of data; each piece has a visible, permanent location. Until someone gives me a better word or phrase, I am calling this type of system "instantaneously structured data."

Here are a few of these "unthinkable" things that instantaneously structured data might do:
> Researchers have yet to develop a theory of measurement – and the choice of wrong measures can cause managers to make unfortunate decisions. For example, gross margin percentage is a widely used metric

of success. It drives managers to get out of the low end of their product markets, and prioritize products whose gross margins are highest. It causes companies to be disrupted. If managers choose another metric – like net profit per ton or per car – it creates in many instances a drive to stay at the bottom of a market to prevent disruption. An instantaneously structured data system would allow an executive to instantaneously develop income statements that measure success in different ways.

› I can imagine using instantaneously structured data to create five different organization charts and use them simultaneously to show employees how they respond to different executives for different elements of a project; it would allow an executive to instantly convene and then disband members in teams of various members, and it could organize a company around the system for developing new products.

› Finally, for the first time I can see how an organization can be structured not around my company's economics, nor around the customer. Rather, I can envision the organization focusing around the job that the customer is trying do get done. Because jobs to be done are stable over very longtime horizons, I can imagine that with instantaneously structured data, never again will executives need to reorganize their companies.

I thank Hasso Plattner and SAP's employees for developing this technology and making it available commercially to all. By analog, when IBM introduced its Computer 360 in 1964 those of us in that generation were awed at the technological achievements that enabled the machine to do its magic. The next generation felt the same when IBM introduced their PC. Technology that gave computing to average people truly was viewed as a miracle. With the passage of time, however, I increasingly admire the managerial achievements in the 360 mainframe computer and the personal computer. The work of tens of thousands of people and billions of dollars had to be orchestrated within and between massive, integrated companies and small entrepreneurs. We were excited about the few applications that were known for these generations of computers. But we were even more excited about the many applications that had not been conceived.

For the same reasons, I stand in awe of the technology from SAP that enables instantaneously structured data. I admire their engineering prowess and their managerial skill. I enjoy thinking about what the technology holds in store for managers and their organizations – things that we cannot now imagine. Instantaneously structured data will not solve the problems of management. But I do believe that this technology has the potential for putting the problems of speed, bureaucracy, and one-size-fits-all marketing onto the back burner. And for now, it is fun to think about unthinkable things.

Preface

Why do I, Hasso Plattner, write this book together with Bernd Leukert and some of his people at SAP?

Since 2006, I have been working at the HPI at the University of Potsdam on the development of a new foundation for enterprise systems. In the meantime, the scope has grown much wider from healthcare or traffic control, to genomics and Ebola containment – but, the enterprise is still in the focus. A prerequisite for new enterprise systems is an ultrafast database – HANA. With HANA now having matured, SAP can build a new enterprise business suite, S/4HANA.

Many papers and books have already been published. Some are really good, others... Therefore, I feel compelled to tell my story and what I think is important in the future, together with the practical experience of SAP, who does all the hard work of building a new system under the obligation to carry the existing customers forward into S/4HANA without disruption. This requires a new level of engineering quality, and I am honored to have SAP as a partner.

The book should enable people in management to recognize the potential of S/4HANA and to feel confident to start projects, small and large, in the near future. The potential of S/4HANA is huge, with our step by step approach we see a growing impact on Total Cost of Ownership (TCO), user experience, new business processes, and most importantly, a better interaction with the customers. You wouldn't believe my prognosed TCO savings, but here are some facts: data entry is three to four times faster, analytics ten to one thousand times faster, the development of extensions much faster, and there is significantly less database administration work, with a data footprint of 1/10 and an unlimited workload capacity via replication. And on top of this, we have the new user experience, the new possibilities for customization, and a whole suite of new applications.

To move from R/2 to R/3 was huge; the move to S/4HANA has many more benefits.

CHAPTER ONE

RETHINK
TO
INNOVATE

I t's fall 2006; I, Hasso Plattner, am a professor for computer science at the HPI in Potsdam, Germany. My chair has the focus on enterprise system architecture, and I have to find a new research area for my PhD candidates. Yes, they have to find the topic for their dissertation themselves, but I wanted to guide them towards something I was really familiar with, a concept for a new Enterprise Resource Planning (ERP) system. All my professional life I have worked on such systems, and I ask myself, what would they look like if we could start from scratch?

A stack of blank paper, pencils, and an eraser – that's all I need for the start, hoping that new ideas would just come streaming out of my head. They don't, and I take a little tour through the Internet with the help of Google and Wikipedia. How the world of computers will look five years from now is the goal of my fact-finding mission.

At the Massachusetts Institute of Technology (MIT) in Boston, they said we might have computers with only marginally faster CPUs, yet many of them. The multi-core architecture was up and coming, and systems with up to 1000 cores were predicted. And the directly accessible memory, whether as DRAM (dynamic random-access memory) or SSD (solid state drive), will increase 100-fold and become cheaper and cheaper. These are encouraging facts. At SAP, we already have a database running completely in memory and working as an accelerator for SAP's Business Warehouse – the Business Warehouse Accelerator (BWA). It is only a small step to conclude that the new ERP system has to run on an in-memory database. But what does this mean for the architecture of ERP systems?

Ever since we have been building such systems, first at IBM, and now at SAP, they were based on the idea that we know exactly what the users want to know. In order to answer their questions in a reasonable time frame, we maintained aggregated data in real time – meaning that whenever we recorded a business transaction we updated all impacted totals. Therefore, the system was ready to give answers to any foreseeable question, thus labeled a real-time system. The new idea I come up with is to drop these totals completely, and to just compress the transaction data per day while keeping all additional data intact. It is not much, one piece of paper, but it is a start.

With this, I went to my team of PhD candidates and educated them about data structures and data volumes in typical ERP systems. After a lengthy session on the whiteboard, one student asked me what the compression rate might be from the transaction data to the compressed data. I did a calculation for a fictitious financial system, and after a while, came up with the answer 10:1. Figure 1.1 shows the original whiteboard with the calculations for data volumes and data access.

The student wasn't the least bit impressed, and said, "From an academic point of view, this compression rate is not very impressive." My new idea, all that I had, was shattered. I took the eraser, wiped out the aggregates in the drawing on the whiteboard, and replied, "Okay, so no aggregates anymore."

This was the breakthrough that I was looking for. No one had ever built a financial system without materialized aggregates, whether updated in real time or through batch processes.

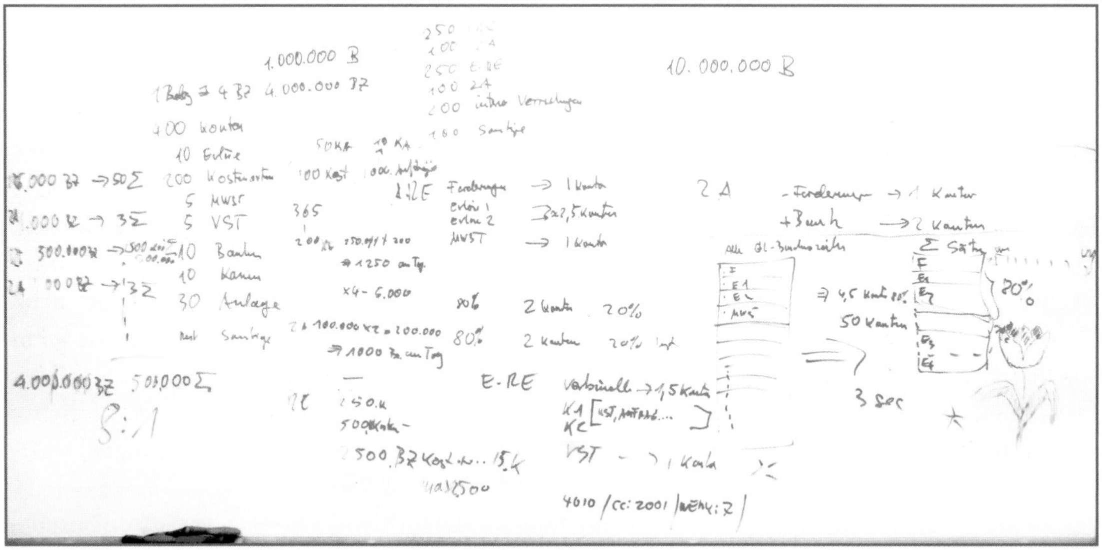

FIGURE 1.1
The original whiteboard at the HPI, where the story of HANA began

Back on the offensive, I asked, "What if we assume the database always has zero response time, what would an ERP system look like?" This was a proper research question, and the academic work could begin. But how about some experimental work? Shouldn't we check for a database that could come close to this ideal? Was this, in the end, possible at all? This is the wonderful part of doing research at a university. At SAP, ideas such as a zero response time database would not have been widely accepted. At a university you can dream, at least for a while. As long as you can produce meaningful papers, things are basically alright.

We asked SAP whether we could have access to the technologies behind their three databases: TREX, an in-memory database with columnar storage, P*Time, an in-memory database with row storage, and MaxDB, SAP's relational database. My PhD candidates started playing with these systems, and it became clear in a very short time that building a new ERP system was, by far, not as interesting as building a new database. In the end, they were all computer scientists – accounting, sales, purchasing, or human resource management are more the scope of a business school student. The compromise was that we build a database prototype from scratch, and all the application scenarios with which we were going to verify the concept of a superfast database had to match those of real enterprise systems.

The reasoning is simple, with zero response time in the database we don't need any of the constructs that we find in all enterprise systems to guarantee a decent user response time.

The aggregates were gone already, next came the redundant tables explaining how the aggregates were built up, all caches in the application, and all the tricks with database indices. We just keep the transaction data, and everything else will be calculated on demand, again and again. Gone were the asynchronous insert/update tasks I had introduced in 1970 at Imperial Chemical Industries (ICI), then still working for IBM. Gone was all the unnecessary locking of database objects around updates. The new application code looked clean, simple, and short. The database footprint would be much smaller, and a reduction by a factor greater than ten seemed to be achievable which was very helpful considering that we wanted to keep all data in-memory. Our database prototype could do only some of the required functions, but it was clear from day one that both Online Transaction Processing (OLTP) and Online Analytical Processing (OLAP) should run on one single system. We will later see that this is the foundation of a new approach for building enterprise applications.

Why did nobody else think about a database with close to zero response time for transaction and analytical processing? Probably we were focused so much on direct access via indices that we couldn't imagine using columnar storage in high volume transaction processing, or we simply couldn't think about a management information system without totals in multidimensional cubes.

We got anonymized real customer data, from a small German brewery and from a very large American Consumer Packaged Good (CPG) company, to experiment with. The Profit and Loss (P&L) statement for the brewery, running through every single anonymized accounting line item, took only three seconds. The dunning process for the large CPG company – at first, three seconds, and later, less than one second. We got bolder, and tried to build an Available-to-Promise (ATP) system without any preaggregated totals, only to find out that it is possible and that it has huge advantages in solving potential conflicts.

Everything that we had learned after two years of research was published in a paper that I presented at the SIGMOD database conference in Providence, Rhode Island in 2009 (see Figure 1.2) [Pla09]. The representatives of the established database manufacturers were good listeners. All of them started in-memory initiatives shortly after the conference for their major products. Michael Stonebraker, an MIT professor, asked me how SAP could ever rewrite 400 million lines of code. I answered that yes, that is the size of the SAP Business Suite. SAP has to find a way to rewrite every bit of it.

1.1

Starting from Scratch

SAP had listened to the HPI, and started to develop a professional version of an in-memory database with columnar storage for OLTP and OLAP workloads – HANA. It is a huge step from a research system to a production system. Just think about the ACID (Atomicity, Consistency, Isolation, Durability) properties of a relational database, high availability, compatibility with incumbent leaders, infrastructure for application development – to name just a few. Some database experts in the world criticized our

research and SAP's bold move. But this gave us the confidence to continue on the track. Perhaps, now in hindsight, it makes even more sense that SAP acquired Sybase, a seasoned database company with additional database experts.

Before I could see my ERP system dream become reality, the database had to mature, and SAP used the early versions as the basis for its data warehouse product SAP Business Warehouse (BW). Since mid-2012, SAP has been rewriting its enterprise applications (ERP, CRM, SCM, SRM, and PLM), and a first version with a dramatically reduced data model, massively improved performance, and new application possibilities is now shipping. Human Capital Management (HCM) is merged with the SuccessFactors applications and is therefore another project. All ERP systems from SAP ERP Central Component 6.0 (ECC 6.0) onward can be upgraded without disruption, which is a miracle in

Presentation

Hardware Trends

Vision: Boardroom of the Future

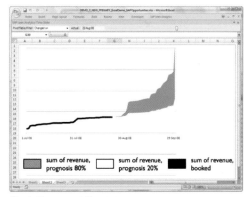

Cash Forecast Prototype

FIGURE 1.2

Impressions from the SIGMOD '09 database conference, where the research results behind HANA and the vision of analytical processing on a transaction system were presented for the first time

itself but easy to grasp when you fully understand the power of SQL (Structured Query Language). What was yesterday an aggregate table became today a view of the transaction line items, likewise a redundant line item table became a projection on the same transaction line items.

That this system is not only much faster in reporting and analytics because of the in-memory columnar storage, but also has transaction processing, which runs three to four times faster thanks to the removed redundancies, was not expected by the experts. Many times we were asked what the business advantage of the increased database speed is. The answer is very simple; we can now concentrate on the business logic and not spend the majority of our time on performance enhancing constructs. Finally, we can deliver on the promise we made all these years to build a real-time application system for enterprise. With the simplified applications we can do so much more, and it is worth having a look at the development in more detail.

By reimagining and starting fresh, we have begun the in-memory revolution with SAP HANA.

1.2
A More Theoretical View of Enterprise Applications

When we refer to enterprise applications, we mean applications in sales, planning, order fulfillment, manufacturing, human resources, purchasing, transportation, research and development, call centers, administration, and performance management. Not all areas are covered by standard products from SAP as of today. Therefore, with a new database and new applications we will not only improve existing products, but also open up new markets.

Enterprise systems record a lot of data even in smaller companies, and for the larger ones this data approaches petabytes. Most of the data is structured and entered via transactions in which the input is verified and added to the database. This part is very technical to guarantee the correctness of the data according to the data model and specific business rules. Typically, these transactions deal with business objects such as orders, payments, shipments, etc. All attributes of a business object have to be valid; therefore, they are checked against hundreds of definitions of codes such as currency, country, measure, date, tax, etc. All references to other business objects have to be valid as well, and some basic rules for business objects need to be fulfilled.

The real application starts when we work with these business objects, e.g., when we prepare orders for shipment, invoice customers, or reconcile incoming payments against open debtor items. For some of these applications performance is already a concern. The faster we can execute these processes, the better for the general flow of work in a company. A different type of application is planning; wherever possible, we try to have a plan. Manufacturing, sales, and financials all rely on planned figures to measure the actual business against the plans and control the performance of the company. The majority of applications are of a reporting or analytical nature. With these, companies want to gain insights into the ongoing business, document for legal reasons, analyze correlations, detect deficiencies, and monitor emergencies.

All of these applications have to deal with a large amount of detailed data – the transactional facts – and then try to convert this data into meaningful information for the people in charge. Since people are only able to work on a relatively small amount of information at a time, these applications have to transform the details with the help of aggregation, averaging, weighing, etc into a more comprehensible portion of data. If we assume that people can consume up to 12,000 characters in one report (10 pages × 30 lines × 40 characters), then we deal with a compression by a factor of several million (from gigabytes of raw data to ten kilobytes of one report). It is obvious that information technology has tried really hard to anticipate the future as much as possible by preparing multiple layers of compression in management information systems. For this reason, we used materialized aggregates, redundant data, and triggers – all being redundant transformations of the original transaction data – to understand what is happening.

To get rid of all this derived data and replace it with on-demand algorithms, only executed when needed and not in advance, seems to be a radical idea. Instead of duplicating data, we create temporary datasets for immediate consumption. If successful, it will drive a tremendous simplification in the way enterprise systems work. It was only a small step in the early discussions from dropping aggregates to a completely redundancy-free application system. Instead of preparing information in advance and storing it as materialized aggregates in the database for rapid consumption, we developed a cascade of applications presenting information with various layers of detail. Having the freedom to basically switch from one layer to the next, going back and forth (thus having full freedom of navigation) from dashboards with key figures or graphic displays to multiple levels of aggregation, while applying filters and other controls directly on transaction data in real time, is the foundation for a deeper understanding of every business and sets the ground for promising new applications.

As another dimension to these innovations, unstructured and structured data can be processed in one system. There is much potential for unstructured or partially structured data in an enterprise. Future applications will include all kinds of messages, product descriptions, pictures, video, audio, etc. Even if these data sources reside outside the database, it can index the data and refer to the originals when needed.

Various application types, e.g., OLTP, OLAP, and Business Intelligence (BI), can be combined to run in one system, making the workflow much more integrated and real-time. The applications will become more intelligent, and move from the simple record-execute-document mode to a supporting record-schedule-predict-advise-execute-learn mode. Since we don't have to worry about performance so much, we can spend more time on intelligence. To hide the complexity from the user wherever possible, all of this requires a new style of user interaction. Instead of data entry (OLTP), automated actions workflow (OLAP), and reporting (BI), we need multifunctional applications with built-in analytics, bringing various data sources (including external ones) together and enabling immediate action in case of necessary corrections or decisions.

For example, the advanced ATP logic works on line item level, the finest granularity, instead of aggregates for stock, planned production, and customer commitments. In case there is a conflict with regards to volume or date, the application analyzes the customer situations, simulates different schedules, and makes proposals for the best possible solution of the conflict. Only working on the highest level of detail allows for this new intelligent behavior of applications.

1.3
About This Book

Our journey in writing this book started with the simple question: why is it so difficult to communicate the benefits of HANA and HANA-based solutions? We found that, too often, customers don't know how HANA enables innovation and how it generates business value. As a consequence, they don't realize that HANA can be the solution to their individual business problems.

Although there already exist a number of publications about HANA, they are often written from (and limited to) a specific point of view. While existing literature on in-memory columnar databases [Pla13] focuses on technical readers, a rapidly growing number of publications on the success of HANA-based enterprise applications address business people (see [SAP14b] and references therein). This separation makes it difficult to get the whole picture of HANA's technical innovations, and how it enables businesses to run better.

In this book, we close this gap and show how HANA's technological achievements and business benefits are two sides of the same coin. We demonstrate the features of HANA with examples of real applications we have built together with our partners. For each of these examples, we discuss how HANA's technology enables the respective business benefits. For instance, we explain how HANA stores data at its finest level of granularity, and how this enables a new, previously unknown kind of flexibility in enterprise applications.

Structure

This book contains three parts, each coded with a different color.

Part I

introduces the foundation of HANA and illuminates its technical background.

Part II

discusses the SAP Business Suite 4 SAP HANA and its applications to the world of enterprise business.

Part III

describes examples of real-world HANA projects and explains how HANA enables the particular solutions and their impact on business.

Target Audience

As this example-driven book describes both HANA's business benefits and its technology, we address readers from both business and technological backgrounds. We target executives, consultants, sales people, and startup entrepreneurs, as well as engineers, scholars, and students in the area of computer science. Our goal is to give readers a better understanding of what to expect for the future of business and how they can participate in evolving existing enterprise applications with new technology.

How to Read This Book

While we recommend everyone reads the entire book, we have structured the parts so that readers can identify role-related sections of interest for quick and convenient reference. Each part of the book is self-contained and presents a perspective from either the business or technology domain. Part I is a technical exploration of the foundations of HANA, while Part II presents the second side of the coin, the business values which arise from the presented technologies. Part III combines both business and technical aspects in order to foster comprehensive insights for readers of all backgrounds.

PART ONE

FOUNDATIONS OF IN-MEMORY TECHNOLOGY

Some of the most important developments of the recent years in computer technology are multi-core CPUs (with one CPU currently hosting up to 15 fully functional processing units) and an increase in memory capacity based on a 64-bit architecture, which easily supports terabytes of directly addressable space. Multi-core architecture allows for massively parallel processing of database operations, and since all relevant data is permanently kept in memory, the processing takes place at the highest possible speed. Read operations are completely independent of any access to slower disk storage devices. Write operations also take place in memory, but have to be recorded on non-volatile storage as well in order to guarantee data persistency.

HANA stores data either in columnar or in row format. The focus is clearly on the columnar format, and special arrangements were made to support both transaction and analytical processing on the same data representation. The interface language is SQL and HANA supports data recovery, high availability, and the ACID (Atomicity, Consistency, Isolation, Durability) properties. In the columnar format, the data is highly compressed to reduce the data footprint in memory. The very high data aggregation speed enables us to use radically simplified data models, which again helps to further reduce the data footprint. Today, we see no limitations for keeping all data of the Business Suite completely in memory, even for the largest companies in the world. This has a major impact on the architecture of the consuming applications, the workflow between these applications, and the scope of individual systems, as well as human interaction with these systems.

To have the opportunity to design and build a new database from scratch, without any burden to take care of an installed database, is a rare situation. We are very grateful to be able to avoid the innovator's dilemma, granting us the freedom to be as radical as possible within the boundaries of the relational theory [Pla13]. Many features of HANA go beyond traditional relational databases, yet without violating their basic concepts.

It is necessary to understand the disruptiveness of the technology to fully grasp its impact on today's enterprise systems and the potential for future applications. The combination of speed, the small data footprint, and the simplicity of the data model gives us an unprecedented opportunity to rethink and rebuild enterprise applications. What was yesterday only possible in specialized applications on different platforms, with mostly duplicated data and often limited numbers of users, can now become part of the core enterprise system.

We will talk a lot about speed in this book. We believe that speed is the prerequisite for intelligence and we want to make our applications more intelligent. Instead of just dealing with a huge amount of data in a largely automated fashion, we develop insights, predict short and mid-term developments, get advice where possible, and build a system which becomes our partner in real time. We firmly believe that one of the biggest obstacles in dealing with enterprise applications is their lack of speed. Any slowdown of employees working with an enterprise application has an impact on the performance of the company and can easily frustrate the people. The ultimate goal has to be the real-time enterprise. Data entry, business processes, analytical insights, simulation of results, period end closing – all the tasks supported by a computer system – should perform close to instantaneously. In this way, the system will become a real partner and help improve the performance of a company. Furthermore, too many people complain about the complexity of enterprise systems as a major stumbling block for future improvements. Therefore, the reduction of complexity has a top priority in the design of future systems. HANA has shown in many projects that the database and the application platform play a vital role in achieving these goals – speed and simplicity.

From day one, the spectrum for HANA-based applications was not limited to enterprise systems. Other areas including research, healthcare, traffic control, etc were also in the focus. Balancing the application spectrum helps prevent favoring one type of application over the other.

Outline of this Part

CHAPTER 2 THE DESIGN PRINCIPLES OF AN IN-MEMORY COLUMNAR STORAGE DATABASE
In-memory columnar storage is the technological basis of HANA. Its design facilitates massively parallel processing and the use of Virtual Data Models (VDMs). The resulting speed and flexibility is complemented by advanced features such as data tiering, aggregate caching, text processing, and predictive analytics.

CHAPTER 3 THE IMPACT OF HANA ON THE DESIGN OF ENTERPRISE APPLICATIONS The greatest impact of HANA is that for the first time, transaction and analytical processing can run on a single system. Its performance also allows us to remove all the redundant data structures which had stored results from predefined aggregates. This results in an unprecedented flexibility and a massive reduction in the data footprint.

CHAPTER TWO

THE DESIGN PRINCIPLES OF AN IN-MEMORY COLUMNAR STORAGE DATABASE

In enterprise applications, we structure data in the form of business objects. We distinguish between master data, transaction data, event data, and configuration data. All business objects can be identified by a primary key. Typical master data objects are companies, countries, customers, vendors, products, materials, and equipment, but also accounts or financial assets. These objects remain in the system for long periods of time and change only occasionally. They are not allowed to carry any aggregated information, such as turnovers, balances, or stock quantities, which can be calculated from transaction objects. Transaction objects are customer orders, vendor orders, shipments, stock movements, payments, etc. These objects describe business transactions taking place between master objects. A more detailed description of the different data classes in the context of enterprise applications will be given in Section 3.1: Objects and Relations.

Most objects are hierarchically structured and each node of the hierarchy is stored in a separate table. The structure of an object is defined as metadata in a data dictionary, which is shared between the application and the database. For references between objects, we use their identifiers (primary keys) as foreign keys. Secondary indices, which connect logical objects with each other or with certain actions, will be eliminated. These are some of the basic design principles of relational databases which have been the foundation of business applications for the past 25 years.

The database systems of the past were designed for storing data on disk and enhancing performance through caching data in main memory. As long as main access to data is realized directly via the primary key and the workload is insert and update-heavy, this concept worked extremely well. The price we paid for good insert rates is relatively slow performance when accessing a large amount of objects, or specific parts of them, in a sequential manner. With the help of many indices, database systems tried to overcome this situation. While transaction workloads could be handled well, analytical workloads suffered tremendously and forced us to set up separate systems, called data warehouses, for reporting and analytics. Any separation of systems sharing identical data has huge costs not only for redundant storage, but also for data transfer and synchronization between systems. One of the primary reasons for the development of HANA was to bring these systems, Online Transaction Processing (OLTP) and Online Analytical Processing (OLAP), back together into one unified system. The cost savings will be huge and new requirements for real-time enterprises are pressing for such a move.

Just as jet aircraft have replaced ocean liners and container ships have replaced bulk carriers, thereby changing the transportation business completely, an in-memory database handling both transaction and analytical workloads will change the enterprise application landscape. For the first time, speed, cost, and simplicity point simultaneously in a favorable direction.

HANA stores data in tables – two dimensional structures comprised of rows and columns. We can arrange the attributes of a table either in the traditionally used row format or in columnar format (see Figure 2.1). In the row format, all attributes of a single table entry are stored next to each other as one sequence, using one or multiple memory blocks, and the table entries are stored sequentially. In the columnar format, all values for one attribute of a table are stored as a vector, using multiple memory blocks, and all the attribute vectors of a table are stored sequentially. Mathematically, both formats are comparable, but their technical implementations on main memory have very different qualities. Organizing the values in the form of an attribute vector allows for easy data compression, e.g., with the help of a dictionary, and also allows for high-speed scanning and filtering. Remember, we want to combine OLTP and OLAP into one system with a unified data representation, instead of storing information redundantly in row and columnar format. This results in a lot of sequential processing where the columnar format has a huge advantage compared to traditional row-oriented disk databases. Together with the option to process in parallel, we can achieve a very high speed for filtering or any type of aggregation – some of the main operations in any analytical workload. The speed is indeed so high that we can drop the idea of transactional preaggregation of data, the basis of information systems for the past decades. And, as mentioned before, additional indices for faster data access are no longer required.

Storage in Row Layout

Column Operation

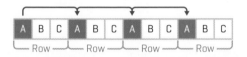

Row Operation

Storage in Columnar Layout

Column Operation

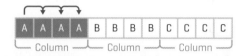

Row Operation

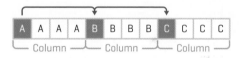

FIGURE 2.1
Row and column operations on row and columnar data layouts

2.1

Core Principles

The decision to store data in a columnar fashion and keep it permanently in memory impacts a variety of design decisions. In the following sections, we will briefly explain the core principles underlying an in-memory database, and then highlight some of the advanced features used in enterprise computing. For more detailed information please see the book "A Course in In-Memory Data Management" [Pla13] or the free In-Memory Data Management course at openHPI.de.

The Superiority of Columnar Storage

Let us look at the two storage formats of databases in more detail. Figure 2.2 shows a subset of synthetic world population data in row and columnar formats. The task is to scan through all people in Germany and calculate the number of males and females. In order to filter out all people in the country Germany using the row format, we have to jump from row to row and check for the country. If a German is found, we evaluate the gender column and update the counts accordingly. If we want to perform the same task using the columnar format, we begin by checking for the

German country code in the country attribute vector. For all identified positions, we access the gender code and then update the counts. The position corresponds with the number of the row.

At first glance, the columnar format seems to be a little bit more complicated. But if we remember that data is always stored in blocks of a certain size, then we realize that for this task we would access far fewer blocks if using the columnar format. With columnar storage, the attribute values are stored next to each other as a vector, allowing for a continuous read. Especially with dictionary encoding, many values can fit into one memory block. The CPU recognizes the sequential process, prefetches the next memory blocks in advance, and keeps them in caches close to the CPU. Since all data is kept in memory and no disk access is taking place, the performance of a task is dominated by the number of cache misses, which indicate blocks that are not in the cache and have to be retrieved from Dynamic Random-Access Memory (DRAM). When accessing the information in row layout, we have to jump in memory once for every person scanned, resulting in 8×10^9 jumps.

For a table in row layout with many attributes, nearly every access to the next value of an attribute leads to a cache miss. On a typical INTEL x86 CPU, the block size is 64 bytes and a cache miss costs 100 nanoseconds [Sco12]. Figure 2.3 shows the different memory layouts for row storage and columnar storage, as well as the accessed memory blocks for two full attribute scans. In row storage, if the accessed attributes cross cache size

Row Storage - Layout

Table: world_population

Columnar Storage - Layout

Table: world_population

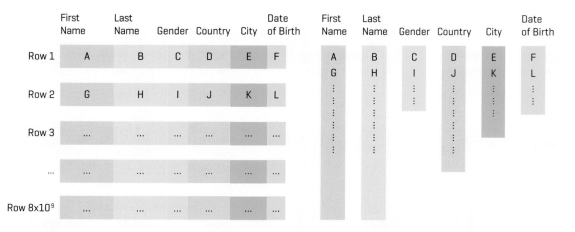

FIGURE 2.2
Comparison between the storage of population data in row and columnar layouts. As first names require more space than gender values, the columns have different lengths.

Row Storage - Multiple Direct Reads

Table: world_population

Columnar Storage - Full Column Scans

Table: world_population

Data not loaded   Data loaded and used   Data loaded but not used

FIGURE 2.3
Accessed memory blocks for full attribute scans of gender and country, in row and columnar layouts

boundaries additional data (in this case, beginning of city attribute) will be loaded although it is not needed. This disadvantage does not occur when using a columnar storage layout.

⇒ **speed is proportional to the number of storage blocks accessed without cache misses**

Dictionary Encoding

With dictionary encoding we can typically reduce the storage needed by a factor of five. Dictionary encoding replaces each distinct value in a column with a unique number representing the value. This encoding reduces the memory consumption as values occurring several times only need to be saved once in full complexity and can be represented by a much smaller integer. Dictionary encoding works especially well for our example world population data (see Figure 2.4) because attributes such as first name (factor of 17), or country (factor of 47), are highly compressible due to their relative low number of distinct values [Pla13]. On the other end, attributes based on timestamps such as a date of birth are less compressible. After encoding, all data in the attribute vector is in integer format, which is the fastest format for a CPU to perform calculations on. As a result, the above task of calculating the number of males and females in Germany runs faster. Despite the compression of data, the distance from value to value is always equal within an attribute vector, and therefore, it is easy to split the scanning, filtering, and aggregation processes into multiple smaller ones, and run these in parallel on different CPU cores (see Figure 2.5). Even in the case of more complex compression algorithms, this remains possible. With these relatively simple optimizations, we achieve a performance improvement of several orders of magnitude for the aggregation of large amounts of transaction items. The important take away is that most database operations work in the compressed integer format, and for the first time, we have parallelism within a single user task.

⇒ **speed is proportional to the number of cores used in parallel**

In columnar storage, attribute vectors do not require memory if they are not populated. This is particularly important for applications serving a broad spectrum of industries in many countries. Such applications with both large amounts and varieties of data would greatly benefit from the ability to reduce the database table size required per customer. Because the speed for aggregations is so high, we can stop preaggregating data. This was necessary in the past to achieve good response times but not anymore. In SAP enterprise systems, we observe a total data footprint reduction by a factor of ten, where compression contributes a reduction by a factor of five, and the elimination of redundant data and aggregates contributes another factor of two.

⇒ **data footprint is about 1/10 of an uncompressed row database**

Columnar Storage without and with Compression

Table: world_population

FIGURE 2.4
Reduction of the memory consumption with enabled dictionary encoding. The compression depends on the number of distinct values per column and reaches a factor of up to 47 in this example [Pla13].

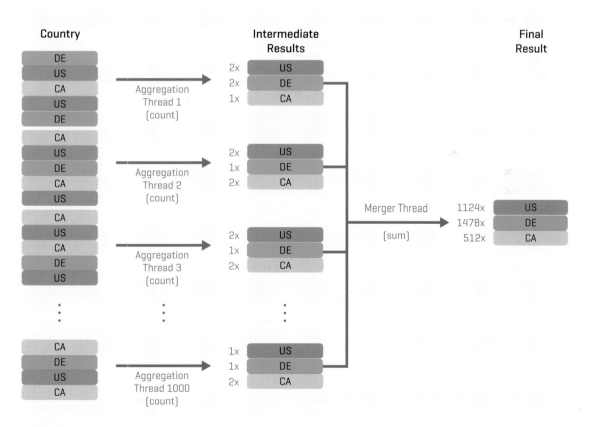

FIGURE 2.5
Parallel aggregation of data on a columnar layout

Large tables, such as Point of Sale (PoS) data, can be split into partitions and these partitions can reside on separate server nodes. With a process called map/reduce we can work on these partitions in parallel, and achieve another acceleration when processing large amounts of data.

⇒ **database operations such as scan, filter, join, and aggregation can run 1,000–10,000 times faster**

The attribute vector scan speed is about three gigabytes per second for a high-end Intel server CPU, e.g., to scan through the entire country attribute vector (8 GB) in our example with eight billion people takes approximately 270 milliseconds if we use tenfold parallelism. Every attribute vector in a columnar database can be used as an index to find specific data entries. We basically only need the primary key, and in some cases, a group key (secondary index) [FSKP12]. This gives the end user a significant increase in flexibility and reduces work for the database administrator.

⇒ **a columnar storage database in memory does not need specific database indices**

⇒ **less database administration**

As we are able to calculate totals by aggregation on-demand and do not post redundant data copies for performance reasons anymore, even the transaction processing speed increases

substantially. The removal of transactionally maintained aggregates also means a sharp drop in database update operations. Update operations (read + write) are not only costly, but have to be handled carefully to avoid data inconsistencies or database locks. In a multi-CPU system, several caches of the CPUs can contain the same data. When an update occurs to this data, a special operation must take place to avoid inconsistent data. This feature is called cache coherence, and becomes more and more complicated with the growing number of CPUs in one server. The columnar database reduces the complex workload for cache coherence in large servers, which helps to avoid frequent slowdowns. The remaining updates to table entries are executed in insert-only mode, meaning that the old entry is invalidated and a new entry is inserted. Any change of a transaction business object remains without further consequences in the database. The changes will only become visible at the time an application program is accessing the changed business object. The simplification in the applications is dramatic, with a huge impact on system quality.

⇒ **update operations are taking place as insert-only operations**

⇒ **database locks are prevented**

⇒ **CPU cache coherence is easier to maintain**

⇒ **application quality improves with simplicity**

The speed of on-the-fly calculations is so high that we can stop preaggregating data.

All attribute values are converted into integer numbers via a dictionary. The integer number corresponds to the position of an attribute value in the dictionary. The internal attribute vector contains only these integers. The dictionary keeps the attribute values sorted; this means we find the integer number in a query for a given attribute value through a binary search. In the opposite direction, coming from the attribute vector we use the integer number to retrieve the external attribute value directly (and the offset is calculated with the integer number multiplied by the respective bit-size). As long as the content of the table does not change, this sorted directory works very well, but new values can change the assignment of attribute values to integers. The HANA database therefore splits the attribute vector into a main vector and a delta vector. The main vector remains stable for a longer period of time, e.g., a day, while all new entries are kept in the delta vector and always added to the end. Since the delta vector is relatively short, the dictionary complexities can be handled much better. From time to time, HANA reorganizes all attribute vectors of a table and merges the delta parts into a new main part. It is this concept

which enables HANA to perform well as a database for both analytical and transaction workloads.

⇒ **new table entries are handled via a delta store for transaction speed**

At this point, it should have become clear that an in-memory database with columnar storage has huge advantages. The fact that every major database manufacturer is adding columnar storage to their row storage is proof of this. Unfortunately for them, this subsequent approach has two disadvantages: first, the lesser compressed row storage does not allow for massive footprint reduction, and second, the inner complexity of the database is significantly higher. Figure 2.6 shows the overall structure of HANA.

2.2

Advanced Features

In the following, we will have a closer look at some advanced features of databases that are of increasing interest for enterprise computing. Some of these features benefit from the columnar layout in particular, other features gain new potential based on the increased speed of an in-memory database.

The Virtual Data Model

After we have dropped all redundant tables, all aggregates, and most database indices, we look at a reasonably simplified physical data model. As already mentioned, every attribute of a table can be used directly or together with other attributes as a database index. With the help of SQL views and Multidimensional Expressions (MDX), we can now define Virtual Data Models (VDMs) on top of the simplified basic data model. The foundation is the ultrahigh scan speed by which we identify subsets of a table and associate them with other tables. These VDMs can become quite complex, but their execution in memory is so fast that no assistance from extra indices or other redundant structures is needed. In contrast to physical structures, these virtual ones cost no storage space and no CPU performance as long they are not being used. Many VDMs (sometimes only slightly different) exist in parallel. A graphical representation helps the developer pick the most appropriate one for a given task.

⇒ **completely flexible data modeling**

These VDMs give application developers a starting point for developing algorithms. The ease of modification or extension of a VDM allows for rapid prototyping as a fundamental principle in program development. Creating complex views does not change any of the physical structure of the database; it is just a programming task. Therefore, there is no need to get the database administrative team involved as developers themselves are able to create views on real data. We will mention it again

Program development should use representative datasets.

and again, program development and testing should take place on really representative datasets. With the help of database performance metrics, we can analyze the data views and constantly optimize them to scale under realistic conditions. The whole development process becomes much faster and the results are already battle-tested with regards to data volumes and runtimes. Figure 2.7 shows examples of VDMs on top of the physical data model.

Parallel Data Processing and New Algorithms

In all these years of online enterprise systems using hardware systems with reasonable parallelism, we ran our applications in single threading mode. For each user transaction, we used only one thread at a time. This was acceptable for mainly short transaction steps with direct data access. To manage parallelism in applications is difficult, and only the database concept of columnar storage has brought us the opportunity to split certain database operations into multiple parallel processes. The benefit is

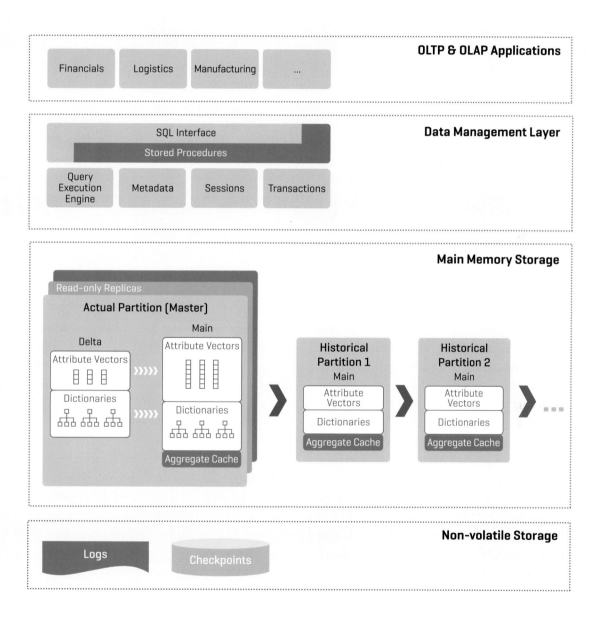

FIGURE 2.6
Overview of the HANA architecture

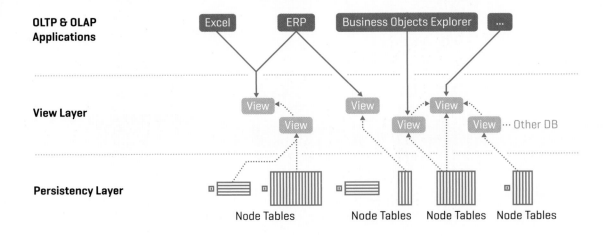

OLTP & OLAP Applications

Excel ERP Business Objects Explorer ...

View Layer

View View View

View View View ⋯ Other DB

Persistency Layer

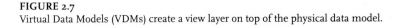

Node Tables   Node Tables   Node Tables   Node Tables

FIGURE 2.7
Virtual Data Models (VDMs) create a view layer on top of the physical data model.

obvious; more complex algorithms and reports for larger amounts of data will run much faster. We can now trade CPU capacity for shorter response times. The trend to have more and more processing cores per CPU will continue, and since the clock rate of the CPUs will not increase significantly in the next few years, parallelism will become a must for our endeavor to build real-time enterprise systems. To establish parallelism is easier for a columnar database than for a row-oriented one, but the real prerequisite is to have all data being accessed within an operation in memory. These two fundamental changes, multi-threading instead of single-threading and in-memory storage instead of disk storage, enable HANA to show revolutionary potential.

⇒ **data in memory and parallelism in processing enable the HANA revolution**

With higher processing speed and parallelism, completely new algorithm classes are possible in Enterprise Resource Planning (ERP) systems. Since all data is already in memory, the execution time is dominated by cache misses and that means direct data access is much more expensive than sequential data access. Instead of trying to access data elements directly via database indices such as a secondary index or other indices, it is often better to use a scan through an attribute vector, especially when a whole set of data elements is being requested. That all attribute vectors can be used as an index creates a lot of flexibility in programming an algorithm without expensive indexing. The database automatically decides when parallelism or single threading is appropriate. As a general rule, programs should try to use set processing instead of following hierarchical structures. Later we will see some examples of this.

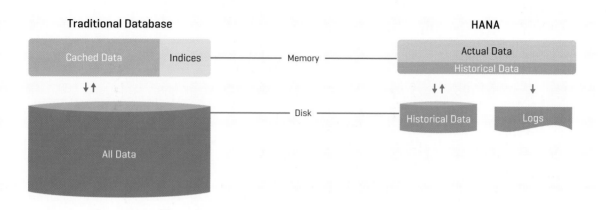

FIGURE 2.8
High-level comparison of memory usage in traditional databases and HANA

Data Tiering

We use the term data tiering for distributing data across multiple storage classes, starting from main memory over SSD to disk. In traditional disk-based databases, we mainly improve the direct access performance with caching algorithms. The most frequently accessed data is kept in the faster storage classes, i.e., Dynamic Random-Access Memory (DRAM), and the lesser used data is kept on disk (see left side of Figure 2.8). With this concept, the database can achieve very high cache hit rates, especially in OLTP systems, and database response times are improved. In a typical ERP system on a traditional disk-based database with row storage, we observe a cache hit rate well beyond 90%. This means nine out of ten data requests can be satisfied accessing the cache and never go to the slower disk. The database

administrator monitors the cache efficiency and adjusts the cache sizes accordingly. With this, all vendors of enterprise systems were able to achieve remarkable response times in systems with high workload.

Actual and Historical Partitioning

Still, another form of data tiering is much more interesting. Remember, we want to keep the majority of the OLTP and OLAP type applications on one single system. In order to make the active data working set as small as possible, we split the data into an actual partition and historical partitions (see right side of Figure 2.8). This is an algorithm based on business status and not on data access statistics. In the actual partition, we store all data which is necessary to conduct the business and needed for documentation for legal and managerial purposes. To the contrary, the historical partition contains data which cannot be changed anymore and is not necessary to conduct the transaction business. The data is, by definition, read-only. The logic for the split is defined in a set of business rules, e.g., in accounting, line items which are from a previous fiscal year and are not managed as open items (reconciliation), can be considered as historical. Figure 2.9 shows common criteria for data to be considered as actual or historical.

In an accounting system, over 90% of all data access is only on actual data. Since we now keep data directly accessible longer (with OLTP and OLAP together in one system), we observe the typical ratio between actual and historical data ranging from 1:4 to 1:10. Most programs work as intended with the actual partition only. This much smaller partition is easy to keep in memory all the time, and many functions like backup and restore, replication for high availability, or data replication for more read-only performance benefit from the split. Only for this active working set of actual data

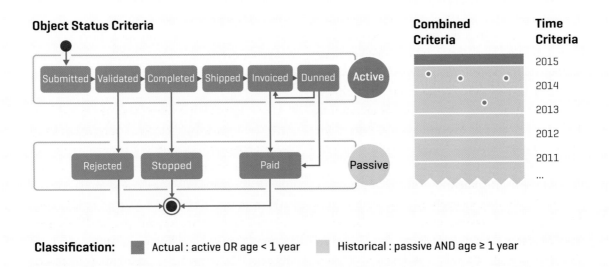

FIGURE 2.9
Criteria for separating data into actual and historical partitions

Applications accessing only the actual data partition will benefit from its smaller size.

do we need the more sophisticated hardware with cache coherence, since only here database inserts and updates occur.

It is easy to understand that all applications accessing only the actual data partition will benefit from the much smaller size. The database workload will be significantly reduced and more users can share the same server node. This also reduces replication efforts, and replicated systems are not only used for high availability, but also for extra capacity for analytics and other read-only processes. Another process, the merging of data from the delta store into the main store, works much faster because of the reduced partition sizes. It is amazing how well a partitioning concept works when the application defines the partitioning rules and the queries carry information about which partitions are needed.

The historical partition does not have to be backed up on a daily basis (actually only after an actual/historical reorganization every quarter or year), and since the data is strictly read-only, we do not need a delta store for new data entries and can apply advanced compression techniques. If we subdivide the historical data by time period (quarter, year), a precise specification of the time frame we are interested in leads to a minimal

dataset to work with. If required, queries for historical data automatically incorporate the most recent data from the actual partition. Whether an item is located in the actual or historical partition depends on its age and its status (see Figure 2.9). Older balance sheet line items, customer orders, vendor orders, etc could still be in use and accordingly be located in the actual data partition. Together with the previously discussed processing speed of in-memory columnar storage, we reach a completely new level for running enterprise systems with respect to both speed and size.

⇒ **splitting transaction data into actual and historical is very efficient as it is based on business status**

⇒ **applications know whether they need all data or only actual data**

Hot and Warm Aging

Another form of data tiering is aging. As we want to keep several years of transaction data directly accessible in our enterprise systems, a distinction into hot and warm partitions is useful. While hot partitions always remain in memory, warm partitions may stay on disk or SSD and come only on request into memory. The format on disk is already columnar, and the load is supported by an intelligent prefetching algorithm (see Figure 2.10). The classification of single entries into hot and warm cannot be based on access history such as in row databases with indices. In those databases, the data access is – to a large extent – direct either via a primary or a secondary index, and access-based partitioning works well.

The new applications on HANA now work much more sequentially: we can use a full attribute scan to identify the items we want to work on, a full attribute scan to build the aggregates on request, a join between related tables, and so on. Therefore, the classification has to be known to the application for it to either request the hot, the warm, or both partitions. The aforementioned actual/historical partitioning should be preferred. Partitioning rules are key for improving performance, and have to be designed carefully by the application architects.

Vertical data tiering with different storage classes in HANA is not always helpful, especially when all partitions are accessed in one query. When handling a huge amount of data, we can reduce main memory attribute vectors which are not requested for a predefined period of time, e.g., a week or month, by temporarily purging those vectors and keeping them only on SSD or disk until there is a new request. Figure 2.11 presents the Least Recently Used (LRU) algorithm that purges unused columns. This works well for data warehouse applications.

Partitioning rules are key for improving performance.

Vertical Data Tiering of Unstructured Data

Unstructured data such as text, pictures, movies, and speech can be managed by vertical data tiering, using different storage classes and caching algorithms. Only the relevant information is extracted and stored in memory, large blocks (BLOBs) containing unstructured data often remain on disk or SSD. A relevant example of this is text search in medical records or scientific publications as described in Section 6.4: Co-Innovation in Action: Medical Research Insights.

Aggregate Caching

When we drop the application-controlled management of aggregates on predefined levels, we definitely speed up the data entry process and gain the flexibility to dynamically aggregate along any hierarchy. But what happens if there are frequent queries for the same aggregate? We do not want to calculate the same result again and again. When HANA detects such a repetitive query, it memorizes the result and reuses it if possible. As long as none of the tables accessed in the query have been reorganized, i.e., during a merge operation, HANA adds the new entries from the delta vectors to the cached result. This is very helpful in cases where the latest table entries have an impact on the business value of the result. A good example for this is the adjustments at the end of an accounting period, or the monitoring of sales data while data is streaming in.

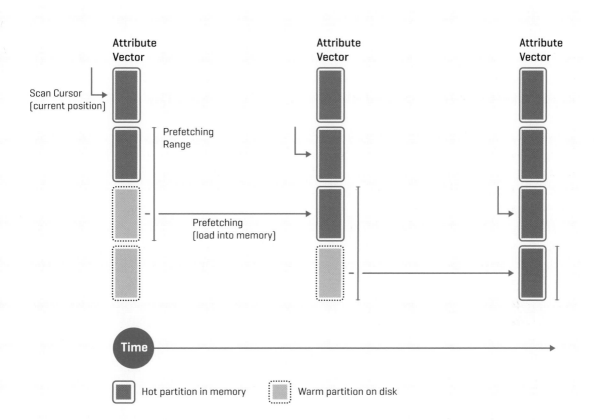

FIGURE 2.10
Prefetching warm partitions and loading them from disk into memory

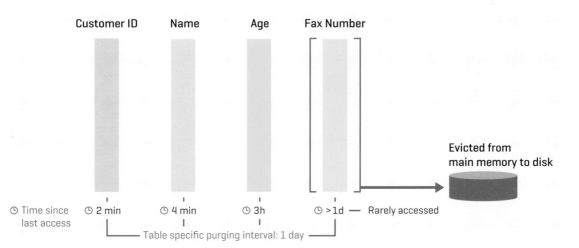

FIGURE 2.11
Purging of columns based on the Least Recently Used (LRU) replacement policy

Figure 2.12 shows a cached result from the past (historical data) combined with a cached result from the main store (only actual data), and the actualization from the delta. The response time is basically defined by the rendering effort in the browser. The same concept works for plan data, with comparison to the previous year and recent modifications to the plan.

When HANA detects a repetitive query, it caches the result and reuses it when possible.

Text Processing

HANA has specific functionalities to process unstructured data such as text. It analyzes text and creates linguistically-aware indices for key words

(see Figure 2.13) defined in a domain-specific dictionary. It also allows for semantic searching, transformations of unstructured into structured information for analytics, correlation with other structured data, and detection of positive or negative sentiments, and stores the results in the database while the text is stored on disk or may remain

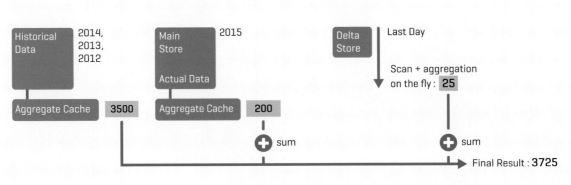

SELECT COUNT ["Invoice_Number"] FROM "Invoices" WHERE "Year" ≥ 2012

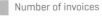 Number of invoices

FIGURE 2.12
Combining the cached results from historical data, main store, and on-the-fly computation of the latest results from the delta store reduces the time necessary for retrieving the final aggregation result.

in a network file system. Via natural language processing, it automatically extracts entities such as people, dates, locations, organizations, products, etc and detects relationships between these entities. The indexed text can then be processed together with structured data in one application. This opens the door for a host of new applications, some of which we will discuss in Part III.

The text search capabilities support full text search and fuzzy search on plain text, PDF, and Word documents. With the HANA Studio development environment, the definition of search models for a web-like query are possible without the need for any extra database indices.

Multi-Version Concurrency Control and Locking

When an analytical query runs, it accesses only the data which was completely processed at the beginning of the query. Transactions (inserting new data or updating data) which are not completed before the query started remain invisible for the analytical query. We call this isolation. The multi-version concurrency control in HANA guarantees this, even if the query runs on

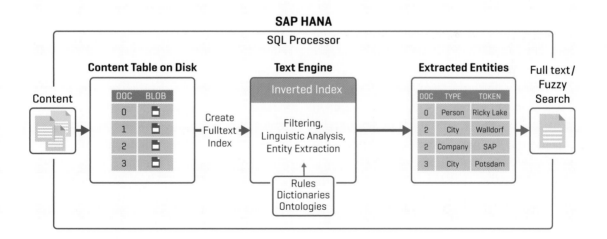

FIGURE 2.13
Text processing in HANA

a replicated system. Many transactions can run in parallel but the database has to avoid two updates happening simultaneously for the same entry in a table. Therefore, the database locks a table entry when the application requests an update, and releases the lock after the update has taken place. After we removed the updates of aggregates in the data entry transactions and only insert data at the line item level, no database locks will occur anymore. As a consequence, these transactions can completely run in parallel without any risk of locking. Now, we must look for remaining potential conflicts. Should two users intend to change the same business object, e.g., a customer's master data, at the same point in time, we have to serialize these two transactions. This could be achieved using the database lock on the customer table entry, but would result in relatively long lock times while the user is executing the intended changes. SAP uses, as an alternative, a logical lock of the object via the enqueue/dequeue mechanism on application level, which ensures that only one user can access an object for intended updates at a time. The database then locks the table entries – only for a very short period of time – during the execution of the update. The actual update in the database is performed as insert-only, meaning that the old table entry is marked invalid and a new entry is inserted. Once the transaction is committed by the database, the logical lock will be released by the application. This solution is extremely helpful when updating large objects with many nodes, stored in one or several different tables.

⇒ **due to the lack of updates most transactions can run in parallel**

⇒ **for transactions with updates the application uses logical locks on business objects**

Replication

The small data footprint for the actual data makes it easy to set up one or more identical replicas. Figure 2.14 presents an overview of the replication mechanism in HANA. All transactions – inserting or updating data – are instantly passed from the master node to the replica(s). Replicas are used for high availability and for increasing the overall performance. All read-only transactions can be executed on the replica without any restrictions. As the maximal potential delay is a few seconds, we do not have any impact on business functions. Database experts might view this potential delay as a problem: does a query run on the latest data, and is there any risk of inconsistency? For read-only queries, there is no risk at all. Whether a query was launched at a given point in time, or one second later, might deliver slightly different results, but this will always be the case. Many business reports run at a certain date, e.g., the end of a month (or day), and are therefore unaffected. Others see the data as it was at the start of the report, and this is not a problem. Only if we take an immediate action based on the data just read, such as a reconciliation, we could run into trouble. Therefore, programs which update or insert data have to run on the primary partition. In addition, they have to prevent other programs from manipulating the same business objects

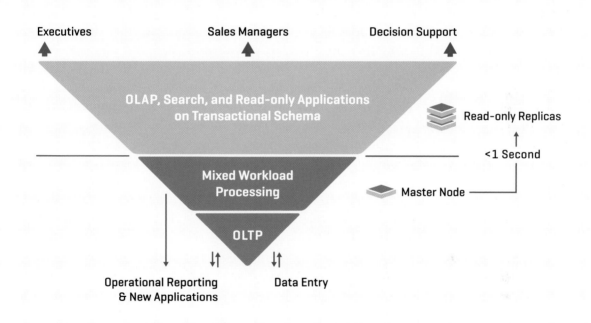

Executives   Sales Managers   Decision Support

OLAP, Search, and Read-only Applications
on Transactional Schema

Read-only Replicas

<1 Second

Mixed Workload
Processing

Master Node

OLTP

Operational Reporting   Data Entry
& New Applications

FIGURE 2.14
Replication of read-only instances supports scalability of HANA systems.

while they run. They have to logically lock the business objects in question. This means, for example, no reconciliation transaction is allowed while a payment program is running. The good news is this locking will take place only for a very short period of time. In case the application is influencing physical systems in real time, it should run on a separate system.

With replication, we enhance the capacity for intelligent analytical applications running directly on the transaction data. Depending on the application, this could be a huge advantage over traditional solutions with data warehouses or data marts, where we always have a certain delay until data becomes available and we have to control the consistency.

It is important to understand that the greatly improved response times for reporting and analytics will create an increased demand, and it would be completely wrong to limit the users again by putting constraints on using the system freely. The overall design of the new Business Suite focuses on simplification and increased use of the system. We firmly believe that the increase in computing capacity decreases the costs of operation and largely improves usability and user efficiency, and is therefore, the right way to go for the future.

⇒ **with replication we increase high availability (hot fail over)**

⇒ **with replication we greatly enhance the workload capacity**

Multi-Tenancy

SAP has a long history with multi-tenancy. Already back in the days of R/2, we offered a version of our ERP system which could support multiple tenants (>500) in one system. The tenants shared the computer, the system software, the database, the application platform, and finally, the applications themselves. These tenants were separated by a tenant code in the database entries. Every data access filters out the tenant-specific data from the other tenants' data. This feature is still available in core parts of the enterprise applications, e.g., financials, and companies can use it for testing.

The cheaper hardware costs of the nineties made multi-tenancy less attractive even in large companies with many ERP systems. With the advent of cloud-based systems delivering Software as a Service (SaaS), the idea of sharing resources to lower costs further and enable continuous maintenance and innovation has led us to revisit the idea of multi-tenancy. Yet, we always have to keep in mind that multi-tenancy is a cost saving measure and has no positive impact on functionality. In the following, we want to portray some of the different approaches to achieve the above goals (see Figure 2.15).

Shared Everything

In the Cloud, we can run a high number of tenants on a large Symmetric Multiprocessing (SMP) computer system using the same application. The tenants are separated with a tenant ID in each tenant-specific data structure. This works well if the application is fairly lean and all tenants are using it in a similar way. Tenant-specific extensions are kept separately in order to allow easy upgrades to the standard system. As long as the ratio of tenants to system is high, we achieve a good load balancing effect, easy on-boarding, and low costs of operation.

⇒ **ratio of tenants to system >100:1**

When the tenant data footprint and user numbers get much larger, sharing tables becomes a disadvantage. To keep the data sorted by tenant ID only works in classic disk-based database systems, and is not the best way to move forward as the cost to overcome the performance impact might

soon outweigh the original cost advantages. It is better to assign the tenants to classes by capacity requirements, and package them in a concise manner on an array of identical systems.

Shared Application

The tenants would still share hardware, system software, and application software, but their data is stored in separate compartments managed by the database. This ensures full data isolation, and requires that the database offers the appropriate multi-compartment data feature. More and more of heavy application logic moves to the database, or takes place in the browser. Sharing the database

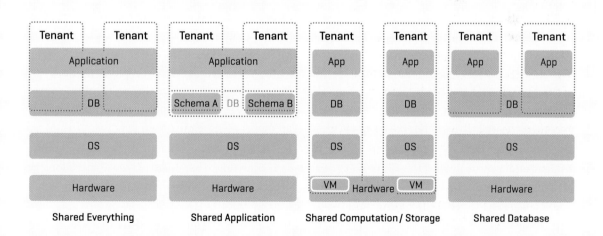

FIGURE 2.15
Comparison of multi-tenancy approaches

and all data center activities is, therefore, more important than sharing the application server logic.

⇒ **ratio of tenants to system <100:1**

Shared Computation / Storage

Each tenant gets its own system, either a part of the physical system with the help of an operating system or as a virtual system. The harmonization of system software, database software, and application software will be established through a coordinated process for maintenance and innovation. This is different to hosting in the Cloud, where each tenant is treated individually.

⇒ **ratio of tenants per system 4:1**

Shared Database

A completely different form of multi-tenancy is to share only the hardware, the system software, and the platform – including a multi-tenant database. Several independent applications get their own data space with parameters for maximum parallelization and memory. With an in-memory database, the data space has to be real and private, while the CPU cores can be shared to some degree. This is an option for larger customers to reduce the number of totally independent systems and to pool management efforts.

⇒ **multiple tenants sharing the database**

Multi-tenancy approaches become much more sophisticated when we consider a modern in-memory database with columnar storage. Why? Simple – the data footprint can be reduced by a factor of ten in comparison to traditional databases. It seems counter-intuitive that we achieve both higher performance for data entry and analytics simultaneously, however, due to the massively simplified data model it is possible. The simplified data model also has a positive impact on the application program complexity which allows for faster innovation cycles.

OLAP, MDX, Predictive Analysis, and R

In general, it is favorable to keep processing of data close to the internal data stores, instead of processing it in the application itself. This is one of the fundamental ideas of relational algebra [Cod90]. This idea was extended to HANA, which incorporates a generic concept of libraries for various business and mathematical functions. First, HANA supports stored procedures – many of them are prebuilt generic business routines for data conversions, date calculations, planning functions, etc. The standard OLAP functions of the SAP Business Warehouse (BW) are now built directly into the OLAP calculation engine. Also, several other statistical functions used in the world of analytics were added. This contributed to the immense acceleration of BW reports when running on HANA.

MDX is a standard defined by Microsoft for handling multidimensional data cubes. These functions are fully integrated with HANA, and allow for an easy integration with the Excel user front-end. Since the aggregations are not materialized anymore, we now have full flexibility to create them on the fly. How much data will temporarily reside in Excel or stays in the database to be called again depends on the application.

Predictive mathematical methods are another important part of modern analytics. A wide range of predictive methods – ranging from classification, cluster analysis, time series, probability distribution, outlier detection, link prediction, statistics functions, and data preparation (like sampling) – is available, directly working on the data in the internal format in memory. Should this not be enough, the integration of the statistical computing software R enables access to an even wider range of mathematical methods used in science and statistics departments. The breakthrough comes with the opportunity to use these very sophisticated methods not only in analytics, but in transaction applications. In the preparation of working sets, or identifying exceptions, we can now be much more efficient.

CHAPTER THREE

THE IMPACT OF HANA ON THE DESIGN OF ENTERPRISE APPLICATIONS

I n this chapter, we will discuss how enterprise systems change when they are designed for HANA. To use HANA instead of a traditional row-oriented database without any changes to the application code is possible, and does result in some performance improvement. But we should not forget that existing databases also keep a large portion of their data in memory through caching. The performance gains using HANA with existing applications are therefore relatively modest, and it is only the five times lower data footprint from dictionary encoding that justifies keeping all data in memory. To make the leap towards a true real-time enterprise – where sensor data (such as from the Internet of Things (IoT)), unstructured data (from social networks, textual and verbal human interactions, and the like), and structured data (from enterprise systems, genomes, etc) come together and can be processed in real time – requires rethinking and redesigning enterprise applications.

⇒ **applications have to become more intelligent**

⇒ **present alternatives for actions**

⇒ **use location, calendar, etc for increased context awareness**

The HANA database works for both Online Transaction Processing (OLTP) and Online Analytical Processing (OLAP) scenarios equally well. To bring both types of systems together opens up a new way of building applications. We will further illustrate this when we discuss specific applications such as Profit and Loss (P&L) analysis in Section 5.2: Financials. HANA is not specifically geared to support SAP's enterprise applications, but these applications will take advantage of HANA's capabilities. There is a wide range of libraries for HANA inside and outside of the classic enterprise space (see Figure 3.1). Columnar storage is generally very advantageous for mathematical algorithms. Most of these algorithms work on large sets of data, and the compact linear structure of the attribute vectors supports data processing. Some of the most spectacular applications on HANA are in this area. Whether it is for hurricane damage forecasting, genomics and proteomics, cancer research, or traffic optimization – the extensive mathematical libraries inside HANA play an important role. Together with the ability to process unstructured data, such as text, and to work on very large datasets, HANA extends its spectrum far beyond the classic enterprise space. In the third part of this book, we will discuss several new applications which exploit the unique features of HANA. In all projects, we observed a remarkable speed of development which is based on the simplicity of HANA, e.g., the removal of indices, no materialized aggregates, and the high speed of execution (especially when dealing with large datasets).

⇒ **HANA is an all-purpose database**

FIGURE 3.1
An overview of HANA libraries

3.1

Objects and Relations

Before we continue to study the impact of HANA on applications, we should take a closer look at objects and relations. In this context, we want to discuss only data structures and not program constructs. The data structures of enterprise applications are kept separated from the programs, and can be used in varying ways by programs written in different programming languages and styles. As already mentioned in Chapter 2, two major classes of data form the backbone of enterprise applications – master data and transaction data. Apart from the two major data classes, there is also event data and configuration data, which – while not making up the main part of Enterprise Resource Planning (ERP) data – holds additional information of importance. A more detailed description of data classes will help us to understand their different requirements, usage characteristics, and effects on application performance. This classification of data is based on business aspects, and not on differentiating technical aspects. Both major object types have a hierarchical structure with one or several nodes. In the Business Suite on HANA we store each node of an object (a segment) in a separate table. A special feature called data guides supports access to all nodes, or a selection of nodes, of an object with one application program call. In most programs, only a small projection of nodes is used, which is well-supported by SQL.

Master Data

Typical master data objects are a company, customer, vendor, product, material, employee, account, etc. Figure 3.2 shows a customer object, its nodes, and their corresponding mapping to tables. All objects are identified by a primary key. In the earlier days of enterprise systems, the database allowed us to connect all nodes of a complex object with pointers (e.g., IBM Information Management System (IMS)). Later, we stored all nodes of an object in one data record

of variable length. The idea was to load the whole object into memory with one data access, where further processing then takes place. In HANA, the data of each node is stored in a separate table, and objects consisting of several nodes can be accessed and combined via SQL.

When we model business objects after their counterparts in the real world, it makes sense to associate a number of methods directly with these objects. The overall structure of an object is defined with metadata; how many nodes, the layout of each

Business Object

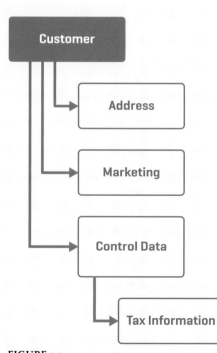

Database Tables

ROOT_ID	OBJECT_DATA_GUIDE
01	[1,0,1,1,...]
...	...

ID	STREET	ZIP	COUNTRY	ROOT_ID
001	Behlert Str.	14467	Germany	01
...	...	...	...	...

ID	CLASSIFICATION	MARKET_ID	ROOT_ID
001	5200	13	01
...	...	...	...

ID	INDUSTRY	TRANSPORT_ZONE	ROOT_ID
0001	6300	10	01
0002	7200	20	01
...	...	...	...

ID	VAT_REG_NR	COUNTRY_CODE	ROOT_ID
00001	...	0001	01
...	...	...	...

FIGURE 3.2
Customer master data and its mapping to tables

node, the definition of attributes including check routines (e.g., for country, currency, language, etc), and the like. We also define which attributes are mandatory and which are optional.

It is important to notice that we do not include any attributes of master data objects which are of a transactional nature. This is a major shift from the past, when we kept many totals in the master data. Whether a customer has, or had orders, is not answered anymore by querying master data totals but by querying the transaction data. A customer master object can be accessed either via the primary key or any attribute of any node, including combinations such as country, city, industry, or others. The result might be a set, meaning that multiple customers were found with the specified attributes. In HANA, all attributes of an object can work as an index, and the user is free to choose which one is most appropriate. Many applications allow for the use of multiple master objects in one transaction simultaneously, e.g., all customers in a given city or all orders due for customers in a certain industry. With the customer ID or a list of customer IDs, we scan the order data. The option to simultaneously look at multiple customers and their transactions opens up a whole new way of working. The master data objects in a standard system carry a large collection of attributes used in many different industries. Most likely, a single company will not use all of these attributes, and it is of additional convenience that HANA does not allocate memory for such unused attributes.

⇒ **master data objects can be identified by any attribute**

Transaction Data

Under transaction data objects, we understand data describing interaction between master data objects, e.g., a customer order specifying products to be sold to a specific customer from a certain part of an organization. In addition, we have customer shipments, supplier orders, goods received, stock movements, financial journal entries, and many more objects. Like the master data objects, the transaction data objects can have a hierarchical structure (Figure 3.3 shows the visualization of a customer order transaction data object). The most complex object is a customer order, which has many nodes, and the simplest one is a Point of Sale (PoS) line item, which has one node.

Object-specific methods are more elaborate. For example, a financial journal entry is checked not only for accuracy of its attributes, but also checked within the whole object for things such as total debit equaling total credit, tax calculation matching General Ledger (G/L) accounts, etc.

For many years, people used the entity relationship model to describe the relation between transaction data objects and master data objects. Technically, foreign keys (attributes containing primary keys) and secondary indices (lists pointing at objects) were used to implement these models. As previously stated, in HANA every attribute column works as an index, and we can group objects by any combination of their attributes on the fly. Therefore, explicit secondary keys are no longer necessary.

As with master data objects, we do not keep any aggregated values in transaction data objects, and instead, substitute the aggregations with proper queries. Keeping data throughout the system on the highest level of granularity is very important, as it reduces dependencies of data as well as the amount of updates, thereby simplifying the entire system. Generic methods, such as "identify all shipments for today," or "identify all overdue invoices," are expressed in SQL as views or stored procedures and become part of the Business Function Library (BFL). Any program, written in any language, can use these methods and SQL views.

Event Data

Event data represents a special class of transaction data. By definition, an event is only written once and is never again updated nor deleted. Event data is only subject to read access after it has been created. From a data structure perspective, an event is often represented as a row in a database table. Event data is typically generated by sensors, that measure certain properties of a target environment or object (such as temperature, longitude/latitude information, pressure, humidity, luminance), or simply identify an object (e.g. RFID).

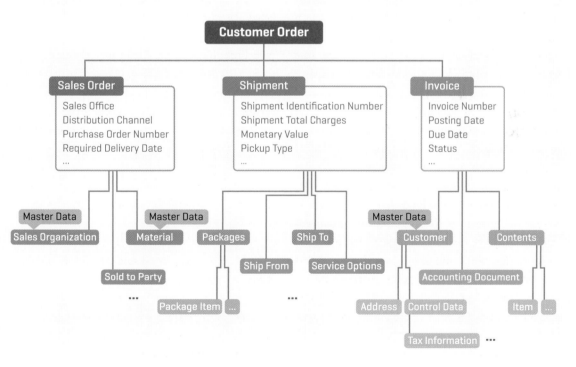

FIGURE 3.3
The hierarchical structure of a customer order transaction data object relates to master data objects in order to describe their interaction.

High sampling rates of individual sensors are common, and a large number of sensors are typically combined into one event stream. Thus, event data is often large just from the number of event rows in a table.

There are sensors everywhere: our homes, cars, aircrafts, phones, cameras, warehouses; even in rovers searching Mars. These sensors either broadcast events over the Internet or store events in local recorders in environments with difficult connectivity. HANA can take all this information, organize it in tables with columnar storage layout, and make it accessible for real-time analysis.

A concrete example where we leverage HANA for event data is a project with the McLaren Formula 1 racing team. Every time a car leaves the pit, a new event stream is recorded. The cars have several hundred sensors, from which over fifteen thousand individual streams of events are emitted. McLaren has been collecting this data for 15 years. When deciding on how the configuration of a car should be adapted between two qualifying runs or for a race, it is extremely helpful to build on experience from the past. For example, racing engineers might need to extract all past runs where the maximum force on the front pushrods exceeded 20,000 Newton. Then, they might want to filter the result to only display those runs where this condition occurred in a corner which is categorized as fast.

Before moving this analysis into HANA, McLaren had attempted to use a data warehouse for the above scenario. In this data warehouse, they stored aggregates for each run (e.g., the minimum force on the pushrods, the maximum, and the mean). This did not succeed because the race engineers had to think about all their possible questions in advance (so that the relevant aggregates could be built up), which limited their flexibility to query the system. As they added more sensors, the data cubes soon grew larger than the original data. Event data, as conventional transaction data, should thus be stored on the highest level of granularity.

Configuration Data

Along with master and transaction data, we have other data for defining codes or parameters for customizing programs. Most of this metadata will become part of customizing objects, and thus, is invisible to the user. Customizing objects present one or more tables in a user-centric form, incorporating explanations for the various parameters. Here, the fact that columnar tables can be extended on the fly helps again to reduce software maintenance efforts.

Applications Using Master Data and Transaction Data

For many years, application program algorithms followed the entity relationship model and its hierarchical structures. We typically began with a master data object and then accessed the related transaction data objects in order to perform a certain task. For example, we read for all customers (in alphabetical sequence) their orders, selected the ones due for shipment, and prepared the shipment. The better way in HANA is to remember how we worked in the days of punch cards. In those days, we would have filtered out the orders due for shipment, sorted them by customer, merged them with the customer master data, and then prepared the shipment. It seems counterproductive, but to follow the mathematical set processing principle is much more efficient than using the hierarchical model. It is all about reducing the total amount of data we access in memory. To use only minimal projections, all the time, is very important for keeping the amount of data accessed as low as possible. With this, the algorithms become much cleaner and therefore easier to write. As a consequence, we have changed direct data access to sequential data access, which is more convenient for columnar storage and is easily processed in parallel.

⇒ **HANA allows for efficient set processing instead of hierarchical program flow**

HANA recognizes where parallelism could be applied and distributes work across the available cores. Dropping most database indices is a huge simplification to the system and the ability to use any attribute as an index gives us real flexibility – especially for custom development or ad-hoc programming. Even for more complex algorithms, there is absolutely no need to get a database administrator involved. The simplification of the data model and its representation in an in-memory columnar storage such as HANA gives us the opportunity to fundamentally rewrite enterprise applications for the first time in many years. With the use of the functional libraries in the HANA platform, the new applications will have significantly less complexity and fewer lines of code.

⇒ **simpler application code**

⇒ **significantly fewer lines of code**

3.2

Redesigning with the Zero Response Time Principle

Some people argue that speed is not a value in itself. We, however, do not share this view. For the past 50 years, we used any available technology to improve the performance of enterprise systems often to only cope with the growing volumes of data and workloads. Speed and accuracy are the very reasons why we use information technology to support business processes. Is not speed the fundamental basis for human intelligence? The ability to reason relies on the number of alternatives our brain can evaluate within a certain amount of time.

The same applies to business questions; the faster we can access data, create a meaningful interpretation, and study alternatives, the better our judgment for actions will be.

⇒ **higher speed leads to more intelligence**

Where would we start with a redesign? We have to put all enterprise applications on the table and ask ourselves how these would look and be organized if the database response times were always zero, or very close to zero. The implications are dramatic; an unbelievable amount of work has been invested into infrastructure and program logic to achieve acceptable response times, but despite this, the business side within a company has in the past mainly felt the negative effects of the lack of performance. The advances in hardware and software now allow us to change this once and forever.

⇒ **no longer any negative impact from the lack of performance**

We have mentioned several times already that all redundant data structures have to vanish. This makes data entry applications multiple times faster and simpler. Many application areas experienced stress from the growing number of transactions, such as the physical warehouse or even accounting systems, where all data from the enterprise comes together. The reduced complexity is important for development speed.

⇒ **reduced complexity saves costs in operation**

⇒ **higher data entry speed means higher data entry capacity**

We learned that database indices are not important any longer, which again reduced complexity, the data footprint, and the workload for the administrators. This also simplified the creation of new reporting or analytical applications. The effort to create a new analytical application solely depends on its mathematical complexity, and not on the data acquisition efforts anymore. The short runtimes for an analytical application enable us to have many iterations tested in a short time frame, and consequently, the whole development process will shorten from weeks to hours.

⇒ **no database administrator involvement necessary for new analytical applications**

The faster we can access data, the better our judgment will be.

HANA automatically massively parallel processes large parts of a report or an analytical application. Theoretically all available cores can be used simultaneously, as long as there are no short running transactional queries waiting. Even in smaller installations it makes sense to set up a replication server and run the read-only queries there.

⇒ **replication increases workload capacity**

OLTP and OLAP applications now easily run in one system, on one database. All reports which were moved to a separate data warehouse only for performance reasons should be brought back. This is another case of reducing complexity and improving the usability of an application. The elimination of the latency until data becomes available for reporting will be more than welcome by the user community. And whenever we have the same data in two or more places, we have to face a reconciliation issue. We believe that some companies will be able to find huge gains in productivity based on these clarifications – just think about planning, scheduling, or optimization activities in manufacturing and transportation. We might even have to rethink some business processes and adjust workflow. More can be done now in real time without delays of work.

⇒ **OLTP and OLAP applications run on one system**

⇒ **huge cost savings and productivity gains**

Most applications have now become short-running transactions and we can chain these together without fearing an interruption of work. This is especially true for the old batch programs. Batch programs are applications which normally take several minutes or even hours to complete. The user traditionally delegates them with the help of a scheduling system to run on their own in parallel to his or her work. The manufacturing industry has made great strides in the past decades to eliminate waiting times and establish a continuous flow of parts and goods. Twenty years after we introduced widespread integration, we now see the potential of a business application world without batch programs. Application programs which produce electronic messages or printed output, such as dunning messages or customer invoices, run as a transaction in seconds while the output is processed asynchronously. Any exceptions can be handled without delay. All that remains is the scheduler with which we can plan for applications to run at certain times or intervals automatically.

⇒ **no longer any need for traditional batch programs**

⇒ **a new business workflow becomes possible**

Since the application-controlled materialized aggregates are gone, any aggregation accesses the transaction data at the highest level of granularity. At first glance this seems to be a problem, but the truth is that any preaggregation created a problem by definition. Using the preaggregated data had locked us into a fixed scheme of hierarchies. In contrast, now we apply a hierarchy on the selected transaction data and calculate the totals on the fly. Any attribute of a transaction table can be used for the definition of a hierarchy.

This gives us the most flexible G/L or sales statistic within the transaction system. We can use many different hierarchies in parallel and the aggregates live only as long as the session lasts. Again, mathematical algorithms instead of procedural programs with transactionally maintained aggregates result in a huge improvement to usage flexibility and system simplification. While simplifying the applications, new features in accounting – such as carve in/carve out in the P&L statement or the balance sheet – and the use of alternative hierarchies and changes in reporting structures become available in parallel to running with the current business setup. The artificial data separation into basic Financial Accounting (FI), flexible G/L accounting, cost center accounting, and US Generally Accepted Accounting Principles (GAAP) vs. International Financial Reporting Standards (IFRS) accounting collapses completely and all subsystems work on the same transaction data. Figure 3.4 presents this simplified data model for Simple Finance [Krü15]. In the past, many applications were used to keep all subsystems in sync. Now, there is only one version of data for a specific business object.

⇒ **no fixed aggregates**

⇒ **no separate subsystems in accounting**

⇒ **no need for reconciliation efforts**

3.3
OLTP and OLAP in One System

In the seventies, we believed the real-time enterprise had become reality. We thought we knew exactly what the users needed in the form of online displays and print-outs, and we trained them to get used to the information systems we had developed. Soon we realized that online reports were only a fraction of the spectrum, and a large portion of reports was still printed on paper. Despite ongoing improvements in computer hardware and networks, we never really got rid of traditional batch processing. Actually, to the contrary, data volumes were outgrowing achievements in technology and information delivery became slower and slower. As a consequence, we decided to separate data into OLTP and OLAP systems, thereby accepting an increase in complexity, storage capacity requirements, and a time delay of up to 24 hours.

The database became the bottleneck even while server hardware exploded in capacity. For OLAP, we introduced a different data model – the star schema – in which the relational tables were denormalized in order to gain speed in analytics. The data came from different systems, and in a preprocessing stage called ETL (Extract-Transform-Load) the data was prepared for fast sequential

Before Simplification

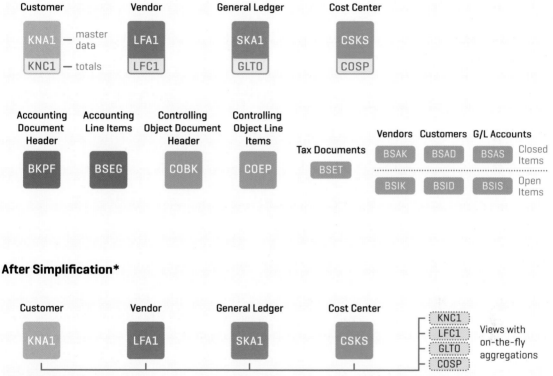

After Simplification*

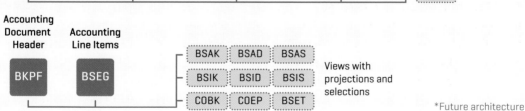

FIGURE 3.4
The data model of SAP Financials, before and after simplification

access. The term data warehouse was born. With the help of various hierarchies, data was rolled up into multi-dimensional cubes. Slicing and dicing – the retrieval of relevant information – took place online and replaced reports on paper. When this concept hit a performance wall again, SAP introduced an in-memory database called TREX and provided it as a data warehouse accelerator. It was always a concept partially created out of the lack of capacity in the transaction system. With HANA, we are back to the seventies with regards to the true real-time enterprise. How HANA does so well in data entry transactions came as a surprise to some people, but that HANA is the perfect tool for OLAP was not a surprise at all. Much of the reporting and analytics functionality can now come back and run directly on the normalized transaction data. The savings in operational costs are huge and the removal of the 24-hour information delay is even more valuable.

⇒ **huge cost savings and up-to-date data**

For many years columnar storage databases have shown their potential, starting from disk-based systems. HANA now brings a new dimension to the OLAP world. The sequential data processing speed is enormous. The only question left regards Dynamic Random-Access Memory (DRAM) capacity and its associated costs. The data compression by a factor of five (for OLAP data) and more already helps. The final solution is enabled through partitioning. As we have learned, HANA can split the data by table and distribute the data across multiple server nodes. Some of the largest data warehouses in the world run now on HANA with unprecedented response times. Most OLAP-specific calculations are now taking place in the HANA kernel in its calculation engine. The distribution of data across multiple partitions and multiple servers allows for a map/reduce process, in which the query is distributed to multiple servers in parallel and on each server the query runs massively parallel again. In large implementations, we are able to let 1,000 cores work on one OLAP query in parallel and collect the partial results for further processing or return to the requester.

⇒ **massive parallelism guarantees very short response times**

A big role in the success of HANA for large-scale analytics has to do with its built-in functional libraries. For enterprise applications, the BFL provides generic routines for the handling of data and currency conversion, year-to-date statistics, decision trees, etc. The OLAP extensions were mentioned already: the algorithms for genomes and proteomes play a role in research and personalized medicine, the geospatial routines (HANA Spatial Engine) for location-related applications, and the predictive mathematics functions and their integration with the statistics functions in R extend SQL standard functions. Figure 3.5 summarizes the HANA platform, its services, libraries, and interfaces.

HANA now brings a new dimension to the OLAP world.

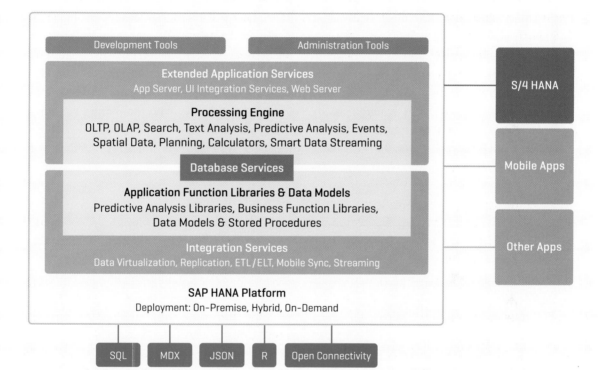

FIGURE 3.5
Overview of services, libraries, and interfaces of the HANA platform

3.4

Time Scaling

The predefined transactional aggregation used in traditional systems did not allow for a fully flexible form of time scaling. Their applications mostly worked on a monthly level. Now, with data access on the highest level of granularity, any form of time scaling is easily possible. The speed of HANA allows queries based on days, weeks, months, quarters, years, or even hours within a day, and trends become more visible. They are the basis for extrapolation into the future. We simply aggregate the individual events into even time scales, and the starting and ending time is completely flexible. These evenly scaled time series are input for a variety of mathematical functions in the statistics library. A graphical display of the data is often more meaningful and will become standard wherever possible. Figure 3.6 presents the time scaling capabilities on P&L data, in which a user focuses on the statements from June and July 2014. Monitoring data over longer periods of time helps us find correlations which then can lead to business assumptions. The

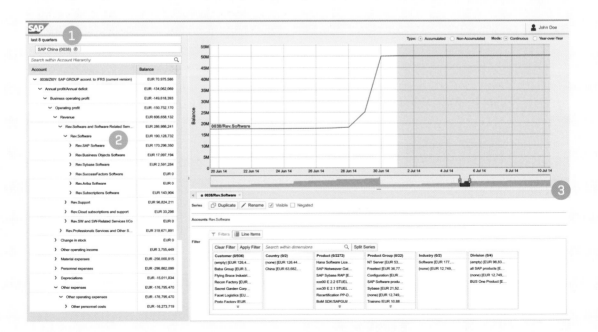

FIGURE 3.6

Time scaling on P&L data. ❶ After selecting an initial time period and account, ❷ users can drill down to the entity of interest. ❸ The corresponding time scale can be further narrowed to a specific timeframe.

high speed of HANA further allows us to work interactively, verifying these assumptions as new insights are developing. Additional techniques such as clustering or predictions play a major role in exploiting the recorded data interactively. To aggregate data into debit and credit per month, as we do in financials, is a primitive subset of the new way to analyze data. The most important fact is that all the mathematics and visualization can take place on the transaction data, inside the transaction system – without any delay.

⇒ **time scaling in reports is completely flexible**

⇒ **sophisticated mathematical statistics are available for transaction data**

3.5

Extensibility on the Fly

One significant difference of columnar storage in comparison to row storage is the fact that for unused attributes of a table, no attribute vector is created. This is especially helpful in standard applications where we typically have a lot of attributes only used in certain countries or specific industries. The designed width of a table (the number of attributes) plays a lesser role in columnar storage. Even more interesting is the option to extend a table on the fly (while the system is running) with new attributes without impacting any of the existing program logic. This plays an important role in the new extensibility architecture of standard enterprise applications. Together with the possibility to easily add fields to the standard screen, it simplifies establishing basic data provision for customer-specific code.

⇒ **new attributes can be added on the fly**

When supporting large enterprises, the need for customer-specific add-ons or completely new applications will be present almost certainly. Using the HANA platform, meeting this need is simplified. The addition of attributes to existing tables and screens is supported and does not impact maintainability. Any major programing language, including SAP's ABAP, can be used to access the added attributes, therefore granting great flexibility. However, in order to achieve the best performance in analytical applications, it is recommended to also rethink existing program flows.

⇒ **customer-specific add-ons are supported by the platform**

It is recommended to rethink existing program flows, as well.

3.6

The Reduction of the Data Footprint

If we use an in-memory database as the foundation for enterprise applications, the reduction of the data footprint is so crucial that we want to have another look at it and discuss the consequences for a system setup. We have learned that HANA compresses data via dictionaries, and that redundant tables such as materialized views (projections and aggregates) as well as most database indices are not needed any longer. This gives us a footprint reduction by a factor of 10. Next, we apply the split of tables for the transaction data into actual and historical partitions. Under the assumption that the ratio is 1:4 (a conservative number), we reduce an uncompressed 100 terabyte ERP system as follows: 100 terabytes gives compressed 10 terabytes, which gives 2 terabytes actual and 8 terabytes historical data. This is a reduction factor of 50 for the transactional part of the system. As mentioned before, only the actual data partition changes while the historical data partition remains unchanged, e.g., for a year. If we apply the same logic to a 1 terabyte ERP system, we arrive at a mere 20 gigabytes for the actual data. These surprisingly small numbers for the actual data footprint have a major impact on the system setup. For a workspace incorporating HANA, the required memory capacity is twice the actual data size.

As long as we stay on a one-node server (chassis, board, blade), we can enjoy fast memory access from all cores of all CPUs. We really should avoid going beyond this setup on one chassis and use multiple nodes for the actual data only when absolutely necessary. Data access across multiple nodes is about three times slower than the average access time for a single node. The largest single-node systems today have up to 6 terabytes of main memory. Incorporating the working space for the database means that actual data sizes of up to 3 terabytes can be handled on one chassis. This is sufficient for the vast majority of today's enterprise systems. In case the actual data size is larger, we use multi-node systems of the Symmetric Multiprocessing (SMP) type which can go up to 12–48 terabytes. We call this scale-up, as we do not leave the single operating system image.

We believe that with this approach, HANA is able to handle the actual data of any enterprise system in the world. It is important to stay on one single operating system in order to avoid unnecessary complexity, such as the previously mentioned cache coherence problem.

⇒ **for most companies the actual data fits on one chassis**

⇒ **an SMP server cluster supports any actual data size**

The small size of the actual data also helps to achieve high availability at much lower costs. Only this data partition has to be considered for backup and recovery procedures. The recovery process will be nearly 50 times faster than for the same system on a row storage database. If we use replication to create an identical copy of the actual data, we not only establish an additional hot standby system, but can also use it for all read-only applications, and thus, increase the performance of the transaction system.

⇒ **the costs for high availability are much lower**

For development and testing we should use the original data even if we have to anonymize certain data. It is sufficient to only use the actual data for developing and testing most aspects of applications.

As described in Section 2.2, historical data should be stored using the scale-out mode once the respective memory capacity is required. The use of smaller server nodes when using scale-out results in a lower Total Cost of Ownership (TCO) without major performance setbacks. Automatic purging of certain parts which were unused for longer periods of time, e.g., a week or more, reduces the memory requirement further if required. Once the data is used again, the data will be loaded from disk before processing begins. The use cases for accessing historical data determine how aggressively we will use purging. For the typical enterprise application we do not recommend introducing other data tiering strategies than the

ones mentioned before. The situation is different for sensor data or when dealing with text and video data.

⇒ **single node or scale-up for actual data**

⇒ **scale-out for historical data**

Splitting the data into actual and historical partitions and using a replica of the actual data leads to a more efficient usage of the computer resources than keeping everything in one system, and is therefore highly recommended. Here, we see a major distinction from traditional databases where the dual data storage, row and columnar, leads to an increase of the original data footprint and hence, is not a long-term solution. In Section 7.1, we present the concrete savings of moving SAP's ERP to the new Business Suite.

The surprisingly small numbers for the actual data footprint inside HANA have a major impact on the system setup.

3.7

Big Data

In recent years, the Hadoop Distributed File System (HDFS) has become a quasi-standard solution for keeping large amounts of structured and unstructured data in a file system. HANA integrates well with this solution and works as an accelerator on top of Hadoop. Not all Big Data has to be kept permanently in memory. Depending on the expected response times, data can be stored completely in memory or partially on disk already in columnar form, from which it is brought into memory on request. With intelligent prefetching, HANA still achieves a reasonable speed for disk-based data that is processed mostly sequentially. In Big Data applications – in which we deal with text, pictures, video, historical, structured, and sensor data – direct data access is rare, and updates almost never happen. With this in mind, HANA can optimize the data footprint and accelerate processing of Big Data. Other important features of HANA are its data federation capability (accessing other databases while processing a query), its data distribution across a large number of servers, and the parallelism of its data services for uploading data into HANA. In Chapter 11, we will take a look at some of the applications of Big Data analytics.

⇒ **HANA integrates well with Hadoop**

Furthermore, HANA's text analysis capabilities enable us to gain insights from unstructured data such as publicly available social media feeds, documents, customer call center messages, etc. We can then share those insights in consumer or commercial transaction systems, where we are in contact with the customer. Section 6.4 and Section 11.5 present examples with respect to HANA's text processing and Big Data.

The geospatial capabilities of HANA range from high speed processing of spatial data to fully integrated visualization techniques, including the use of external Geographic Information Systems (GISs). Section 9.3 highlights these geospatial capabilities for risk analysis during natural disasters.

⇒ **new applications based on text analysis or geospatial data**

PART TWO
NEW ENTERPRISE
APPLICATIONS

I t is March of 2012, and I, Bernd Leukert, am sitting with Hasso and several SAP executives in Hasso's garden in the bay area. HANA has changed the market for database systems and we are debating if the existing SAP Business Suite, with its 400 million lines of code, can be adapted to HANA in a reasonable amount of time or whether we must build a successor to the Business Suite from scratch. From the beginning, it is clear that we have to provide our customers with a non-disruptive technical move from the existing Business Suite to the next Business Suite on HANA. However, how do we introduce such a disruptive technology as HANA without disrupting the business of our customers?

There had been doubt in the organization if this was possible, and some proposed to start instead with a green field, and to create new Line of Business (LoB) applications on top of HANA. I, however, was convinced that we could build a brand new product and, in parallel, launch a HANA version of the existing Business Suite prior to the SAP Field Kick-Off Meeting (FKOM) of January 2013. It was clear to me that if we were not able to recode the 400 million lines of code of the Business Suite, SAP would be missing a vital opportunity to step forward, so we began our work. The plan to achieve our two ambitious goals was as follows:

1. migration of all existing Business Suite systems to HANA
2. optimization of the existing enterprise applications towards the HANA platform
3. simplification by rewriting all enterprise applications

The first step was easy. With a few adaptations, the enterprise applications worked on HANA in the same way they had been working on any traditional database. This was no surprise, since at this point we had not changed any data structures, and we had used the standard SQL interface. The only differences at this stage were a reduced data footprint and an immediate acceleration of analytical queries.

The second step, optimizing the existing enterprise applications, required more consideration. We embedded operational reporting directly into the enterprise applications, bringing transactions and analytics to a single system. HANA became our single source of truth. Unnecessary data transfers, consistency checks, and data reorganizations were immediately removed. This was a big deal, since our experiences with the existing enterprise applications had shown that synchronization with the Business Warehouse (BW) caused up to 35% of the load in production mode.

The third step was ambitious, and aimed at a completely new architecture for enterprise applications through the replacement of aggregates with on-the-fly calculated views. Essentially, we kept the highest and the lowest level of granularity intact. On the highest level, we left all core document tables and configurations as they were before. On the lowest, we provided only virtual views for the aggregates and other redundant data of the replaced system. Subsequently, we eliminated all the data structures that stored redundant data. The result was a huge simplification to the IT infrastructure, as well as to the system itself. Together, this added up to a massive reduction in the data footprint.

In January 2013, we launched the new SAP Business Suite on HANA with over 1,100 HANA-optimized processes, covering all existing components and industry scenarios. General market availability followed in June 2013, and our first goal – the adoption of the existing SAP Business Suite to HANA – was met. We saw the massive data footprint reduction, increased system capacity, and unprecedented system flexibility. Today, we have close to 2,000 customers using the SAP Business Suite on HANA.

In hindsight, the two goals – to adopt the existing Business Suite and build a new one – turned out to be closely aligned. With the Business Suite on HANA, the separation between the transaction and the analytical system was already gone, and the redundant data structures were removed. All that was left to do was to rebuild the existing applications, and to add new ones that would fully exploit the analytical capabilities of HANA. The risk was minimal, as the new enterprise applications would run side by side with the existing ones.

The simplification of the data models allowed a generation change to how we would deal with configuration and testing. We introduced guided configurations which helps the user configure and test new systems in substantially less time. In addition to the data model modifications, SAP Fiori as a new User Experience (UX) paradigm led us to the rebuilding of the application top layer. We enabled new services from the back-end utilizing minimal projections to provide the correct datasets to a web-based User Interface (UI) that is structured around user roles rather than back-end functionality. In total, the revised data model and new application architecture, with guided configuration and a fresh UX, resulted in a completely new Business Suite.

SAP Simple Finance [Krü15] was the first solution that was written completely from scratch for HANA, replacing SAP Financial Accounting & Controlling (FICO). This solution was released at the SAPPHIRE NOW conference in June 2014. Only two months later, in August 2014, we moved our own financial system from the Business Suite on HANA to Simple Finance.

At first, our Chief Financial Officer (CFO) was concerned about the risk. I explained that migrating SAP's financial system to Simple Finance presented a lower risk than any other upgrade. With the support of Hasso, I won the trust of the entire board, including our CFO. At the end of the day, it took us less than 48 hours to move SAP's financial system to Simple Finance and go live.

Simple Finance clearly demonstrated that we were on the right path – it showed a clear impact on both the top and bottom line; we achieved the first two quarterly closings with 30% less processing time and 40% less posting corrections, due to the fact that we could rely on one source of truth. We also saw a 10% Days Sales Outstanding (DSO) reduction in receivables management, and the real-time screening of anomalies resulted in 10% fewer fraud cases, complete with a treasury which could free up 3% more cash through increased business visibility. On the bottom line, Forrester estimates that HANA can save up to 37% of costs across hardware, software, and labor [Par14]. At SAP, we saw a reduction of the database size by 75% only through the introduction of Simple Finance on HANA. Estimations project the possibility of a data footprint reduction of a factor of 14. With this, we are only getting started – logistics, supply chain, production, and customer engagement and commerce all have similar potentials. It is fascinating to see the impact of this technology on businesses.

After R/2, R/3, and ECC 6.0, in February 2015, we launched the new SAP Business Suite 4 SAP HANA (S/4HANA) – the fourth generation of SAP's enterprise systems. The new applications of S/4HANA not only profit from the high analytical speed of the HANA database, but also utilize the nature of HANA as a full-blown application platform. This becomes particularly visible with the new SAP Fiori UX, on which all applications are now built. On the one hand, SAP Fiori is web-based, which means that it runs on any mobile device. On the other hand, all the performance-critical parts of an SAP Fiori application run on HANA, to which SAP Fiori connects directly. All important functionalities are integrated into HANA, and thus available to all SAP Fiori solutions. In the future, the business logic of all enterprise applications, such as Supply Chain Management (SCM), Supplier Relationship Management (SRM), and Product Lifecycle Management (PLM) will run on a single, unified system.

With all this fine-granular, real-time data that HANA provides, we can start to build fully-integrated, predictive applications that help businesses make insight-based decisions with increased accuracy and speed. One of the first examples of such an application with built-in predictive functionality is Material Requirements Planning (MRP), which analyzes different alternatives to solve production shortages and proposes the best solution.

Outline of this Part

CHAPTER 4 **THE NEW BUSINESS SUITE S/4HANA** The HANA platform is the foundation for S/4HANA. It utilizes its speed, simplicity, and flexibility. S/4HANA enterprise applications use the SAP Fiori UX. To guarantee a non-disruptive transition for the customers of SAP, these solutions run side by side with existing enterprise applications.

CHAPTER 5 **REBUILD AND RETHINK** The new enterprise applications of S/4HANA are role-based and introduced by industry. All solutions of the new S/4HANA profit from the improved business features of the HANA platform. This applies to different Lines of Business (LoBs), such as financials, logistics, sales and marketing, and analytics and planning.

CHAPTER FOUR

THE NEW BUSINESS SUITE S/4HANA

After our theoretical discourse, let us now look at the concrete achievements in the development of the new SAP Business Suite 4 SAP HANA (S/4HANA) and how current SAP customers can make the transition. We will then explain the alternative option to move to a full Software as a Service (SaaS) model based on the same code, but configured for a pure SaaS consumer. Taking care of the existing customer base, while applying some of the most revolutionary technology and radical application architecture in S/4HANA, was of paramount importance to SAP's decision making. To effectively illuminate the complex processes at work, we will stress certain concepts throughout the text.

4.1
Non-Disruptive Innovation with Disruptive Technology

We all strive for innovation. We always see an opportunity for improvement, whether it is in a product or a business process. The economy permanently strives for change, and the worst state is stagnation. From time to time, we experience massive technology changes which are disruptive – and sometimes very disruptive.

Think about ocean liners being replaced by jets, bulk freighters by container ships, landline phones by cellular phones, and disk databases by in-memory databases. The replacing technology allows for completely different schedules, costs, and flexibilities – or in the case of databases, new business processes. What do we do with the existing infrastructure? Can it continue to be used? Can it be refurbished? What happened to the ocean liners? They became hotels or cruise ships, but only for a short period of time. Yes, some freighters were converted in the early days of container logistics, but again they did not fit in the long run. And how did our lives change with cellular phones? People do not remember anymore the lines in front of a telephone booth at the airport, or on a busy street.

So, the changes will happen, and they are part of the innovation process. To fight the changes is counterproductive, it costs extra energy and may, most likely, not even work. Despite all this, SAP has talked about "non-disruption" as a strategy in response to the requests of customers.

We promised to carry forward the wealth of data collected over many years, all the proven business processes, including their customizations, and the current User Interface (UI) – the latter one mainly to avoid an abrupt transition to the new system. Other IT companies promised full upward compatibility from 30 years ago – and they have kept their promise. Let us explore some of the possible strategies of today to achieve such compatibility.

We are currently witnessing a number of IT trends which disrupt the way business functions. Applications are offered in a SaaS model, meaning applications run in the Cloud and are offered as a service. The Total Cost of Ownership (TCO) is substantially lower when sharing major components of a system across customers. Especially tenants of smaller sizes can tremendously benefit from this approach.

Perhaps more important is the fact that some individualism has to be forfeited, due to a strict standard of code without modifications. No-touch extensions in a separate system are a possible alternative. This makes sense for applications, where companies do not get a significant competitive advantage from modifying them. The service provider also takes care of most system operations, including maintenance and upgrades, pretty much like the shared services in an apartment building. Completely new are the generic shared services, such as marketplaces, business networks, etc, providing the same set of services to many clients while connecting them as trading partners.

The Internet of Things (IoT) will flood us with data, coming from a myriad of sensors to report the well-being or problems of expensive machinery. What was yesterday already standard for aircraft, we will see in drilling machines or dishwashers tomorrow. We discuss an example of predictive maintenance in Section 11.2: Remote Service and Predictive Maintenance.

Sprawling social networks have become a part of our lives, and as such, they give a testimonial to what we like or do not like, and have become a vital source of information for marketing or business in general. Today, nearly every company operates a call center, where valuable information about products and customers is kept in the form of text. On a similar scale, but hidden within the applications, we see how in-memory databases replace disk-based ones at a rapid pace, with the option to fundamentally rethink and rebuild the applications.

How does SAP play out the non-disruption strategy, when faced by such megatrends? In order to deal with textual data, digest billions of sensor messages, and be able to work as a SaaS application in the Cloud, SAP opted for a completely new platform for its enterprise applications. HANA is not only an in-memory database using columnar storage instead of row storage, but it also offers libraries for business functions, predictive algorithms, single-system Online Transaction Processing (OLTP) and Online Analytical Processing (OLAP) functionality, and distributed data management for data marts or IoT solutions. Technology-wise, HANA is truly disruptive, but that does not mean everything has to instantly change. Let us have a look at the SAP enterprise applications a success story for over 20 years. Thousands of companies have invested billions to set up their systems, maintain these over the years, and develop customer-specific add-ons for competitive advantage.

There is a tremendous business value captured in system configuration and application data. SAP keeps both intact while moving forward from any traditional database to HANA. No data will be lost, and the configuration parameters will remain. Thanks to one of the great standards in IT, the SQL interface, all programs can continue to run unchanged. This guarantees a smooth transition from any traditional database to HANA.

⇒ **all data will be carried forward**

⇒ **all process customizations remain intact**

On the other hand, HANA is disruptive, and the unbelievable speed improvements allow us to drop some concepts of the nineties that were introduced back then to guarantee short response times. All transactionally maintained aggregates could be removed and replaced by views to be compatible, but even their importance is limited in the future. Any kind of aggregation for reporting or analytical purposes is now happening on demand, and many different hierarchies can be used in parallel with varying time scales. Furthermore, the various redundancies in transaction data, with different sorting sequences, are no longer a performance benefit. These data structures are replaced by SQL

projections or views with identical layouts and names. The existing programs continue to run without any further changes. Now, we can drop the redundant data structures and most indices from the database and gain an additional reduction by a factor of two in the overall data footprint.

⇒ **all redundancies have been removed**

New programs will be added and will supersede the existing functionality, but they come in parallel, and as such, continue to support the "non-disruptive" paradigm. A similar approach is being used for the introduction of the new UI. The existing UI remains available, at least for a period of time. The new SAP Fiori applications are being deployed in parallel, and the user or user groups have time to adjust to the new layout and interaction model.

⇒ **the existing UI will coexist for a couple of years to be non-disruptive**

There is a trade-off to this strategy – it takes more time. Yet, it is worth it. All customers have the chance to move gradually forward, keeping major accomplishments of the past unchanged, and in parallel, test the new way of working in the system. The final product keeps two personalities, both the old and the new, in one system. S/4HANA looks and feels fundamentally different, solves problems which were unthinkable yesterday, and is still carrying the business configuration and the enterprise data to the future nearly without changes. There are mandatory changes such as in financials, where the new General

Ledger (G/L) becomes standard and the old one has been terminated, and in logistics, where the infosystem is no longer available. Whenever there are redundant applications, only the latest one remains in the system.

It sounds surprising that the move to the HANA platform is the basis for such advances, but that was always the idea of platforms. They offer services that applications need, and shield them from the ongoing changes in technology. Here, the reduction in the complexity of transaction data has even more dramatic consequences. The ten times smaller data footprint not only allows us to keep all transaction data for five to ten years in memory, but SAP can now reintegrate all the components of the Business Suite into one single system. There were two reasons to split up the enterprise systems into Enterprise Resource Planning (ERP), Customer Relationship Management (CRM), Supplier Relationship Management (SRM), Supply Chain Management (SCM), Product Lifecycle Management (PLM), and Human Capital Management (HCM) transaction systems and to have a separate business data warehouse. First, the sheer size of the systems outgrew single-computer capacities, and therefore, we split them. Second, once SAP had independent subsystems, they could develop them at different speeds using different technologies. Having them all moved to a brand new platform, the HANA platform, invalidates both the size and the speed arguments. All systems can be integrated now, eliminating enormous data transfers between the subsystems. The management of one single system with the above components is easier and

less costly. The separate data warehouse still has value, but much of the operational reporting and analytics functionality can now return to the transaction system. Capacity concerns are no longer a challenge. The separation of actual and historical data, which cannot be changed anymore, and the replication of the actual data partition are the answer, and in addition, contribute to high availability.

⇒ **S/4HANA can run completely in one system with tremendous cost savings**

The transition from SAP ECC 6.0 to S/4HANA is straightforward, without major mandatory functional changes. In the case that the system was already running on HANA, the upgrade effort will be minimal due to the new maintenance process. The stability of the system increases with the drop in the complexity of data entry transactions. The implementation can take place either on-premise, or in the Cloud. Running in the Cloud implies some changes to the maintenance strategy, since the maintenance will be mainly executed by SAP. When running in the Cloud, it becomes much easier to integrate with other generic cloud-based services such as ARIBA, Concur, Fieldglass, SuccessFactors, and many others.

The question is, will everything eventually run in the Cloud? Not exclusively – but it will run there first. There is nothing that prevents cloud software to run on-premise. The financial terms may be different, and the maintenance rhythm will be different, but eventually all innovation will propagate down to the on-premise versions, even if they are based on non-HANA platforms (if technically feasible).

For new customers or existing customers on previous releases who intend to use a green-field approach, the full SaaS model is an alternative. For the SaaS version, SAP took S/4HANA as a starting point, but reduced the needed business objects to only around 30% of the original ones. All compatibility features, such as the previous UI or views representing the old totals, were dropped. The new UI is completely based on SAP Fiori and user profiles (Personas). Customization is supported by customizing objects, which are intended to be self-explanatory. Some of the dropped functionality was redundant, outdated, or specific to industries which are not expected to move to the SaaS version anytime soon. Smaller companies can enjoy the cost benefits of being a tenant in a multi-tenant setup.

⇒ **the SaaS version of S/4HANA is completely stripped of old code**

The SaaS version of S/4HANA will be the first to see most of the future development, and the service provider will take full care of all maintenance and upgrade activities. Despite this version of S/4HANA possessing a completely new code line, all data from previous releases can still be carried forward. For all customers, existing and new, it is mandatory that extensions are built in the no-touch fashion, in a separate instance (see Section 4.8: Extensibility). This is common practice in the SaaS world in order to share as many system components as possible, and reduce

operating costs. Select industries are currently supported, with more to follow based on market demand. All components of S/4HANA are available fully integrated into one platform, and are connected to the generic SaaS applications of SAP or other SaaS service providers.

4.2
One More Time – Speed is Important

Why is speed so important? Especially after years of huge investments, with core business processes finally running smoothly leaving companies to concentrate on new applications currently not provided as standard software offerings, speed is essential.

Why should a running system be replaced now? There is a window of opportunity within which a smarter application system can help to provide the needed information in a timely fashion, and allow the company to make the right decisions at significantly lower costs. And, for the first time in SAP's history, the switch to a new system is non-disruptive for customers using the HANA Enterprise Cloud (HEC), managed by SAP or on-premise. Several other service providers support the HEC model in their data centers.

This offer cannot stand forever because we are again in a storm of change. After the Internet revolution, the challenge of Y2K, the success of mobile devices, and the integration of social media into business applications, we thought that might be it. However, together with the IoT and its sensor data, Global Positioning System (GPS) data, and RFID read events, we produce more and more data volumes – Big Data. The sheer size of this data presents a challenge. However, we believe something else is even more important – the response times of applications still have to be on human terms; to delegate tasks to other people or schedule batch jobs is no longer appropriate.

Why should we compromise in the time of ultra HDTV, video on-demand, and other benefits of the digital revolution? Let us take games for example. Their response times are absolutely crucial, and huge investments are a necessity to let the action happen in real time with high fidelity. Or web search, where we expect to find basically anything within a few seconds of work with the help of a search engine. We flip through large numbers of pictures, watch mini-movies, study interesting topics – all in real time. We are now nearly permanently connected via the Internet, through all kinds of devices, to different applications. The latest trend is to have the latest information at our wrist – the smart watch. Once we experience slow response times, we shake

our heads and just walk away from the current task to do something else. If we talk to someone and do not get a response within a few seconds, we will leave the conversation. This is the same when interacting with a modern computer application. Since the costs for a speedy service are much lower than the costs of people being kept waiting, idling, delegating, starting, and stopping, we have to be on this high-speed, high-fidelity alternative.

The world of computer systems has finally become human. Response times have to be sub-second regardless of how many concurrent users are sharing a system. Only for some more complicated tasks do we tolerate waiting times of up to three seconds without getting annoyed, and after about eight seconds, we start doing something else [Mil68, Bic97].

⇒ **response time categories of one, three, and eight seconds**

Does this behavior really impact our enterprise applications systems? People working on enterprise systems are used to batch processing. They know that certain tasks take minutes or even hours to complete, so they schedule these tasks to run overnight, and continue their work on the next day. Only if the response times during their interactive work become really long, do they start complaining. Yet, sometimes they ask why enterprise systems are so different from the rest of their experiences with web-based consumer applications.

It might be that we are losing a huge amount of human capacity by frustrating our workforce, and that they still come to work should not blind us to their struggles. We have to change soon. If we want to use analytics, planning functions, and optimization runs as dialogues in real time, we must meet the aforementioned response time criteria. The number of parallel threads, or rather, physical cores per node, determines the lowest possible response time for a given query.

In many ways, the speed – or lack of speed – of software applications determines how we organize business processes. Yes, it is great that applications can now handle large amounts of data, but what if we could receive the answers to our questions within a normal conversational time frame? What if we could ask a second question, have a discussion, change some business assumptions, and start all over again? It is the speed of the people in a company that matters. If they can back up their assumptions with data, if they can see trends which were hidden in the standard reports, if they can dig two levels deeper to find the root cause for certain business anomalies, then the probability of managing the company more effectively is high. The more we can understand what is happening in the marketplace, the better we can recognize our shortcomings and our opportunities, and the earlier we can make decisions. The companies which will make computer systems a valuable and active partner in boardroom discussions might be able to outperform their competitors. In these companies, the days for static presentations are numbered, and data analyses in real time will take their place instead.

⇒ **boardroom discussions require instant insights**

Consumers are not waiting. More and more business is taking place over the web. We search for websites, browse through catalogs, watch product videos, and want price and delivery data instantly. We expect a shipment within days with a full refund in the case that we return the goods; look for online instructions on how to use a purchased product; ask a call center for help; request remote assistance for diagnosing a defect; provide photos of a malfunctioning device and expect the repair to take place remotely and instantly. All of this is possible today, and many great solutions are already in place. But now, it will become mandatory for everyone selling on the web to also react in human response times [CM01].

⇒ **consumers on the web do not wait**

A crucial part of modern business is advertising: billboards, print media, radio, television, and now, the Internet. The Internet is a medium through which we can get instant feedback from consumers. Trade promotion management is evolving to take advantage of this instant feedback. A good example of the necessity for speed in the window of opportunity is the bonus calculation per customer at the checkout counter. A company in Japan changed the way they do this customer bonus calculation. Instead of doing it in batch mode (a runtime of more than 12 hours), they now offer the bonus, calculated on the fly, while the customer is still in the store. This changes the conduct of business, and its impact can be huge. Yet, the size of this impact depends on response times of under one second. For many applications, we have missed this window of opportunity in the past. Therefore, we should revisit all business processes in which the applications were not up to the task, and check whether they could be reimplemented on HANA in order to meet the desired threshold.

To exchange large amounts of data in real time between a retailer, a distributor, and a producer is technically no problem anymore. In doing so, all participants have a chance to optimize their logistics processes, and by doing so, save time and money. It is understood that only if the optimization can happen in real time will it be effective and beneficial for all partners.

If we bring together human observations and emotions with data from the IoT, correlate them with structured data such as product construction plans, and search the history of similar cases, then we can have a huge impact on predictive maintenance for machines. Or, in another possible example, we can process genomes to detect specific variations for which we might find correlated clinical patterns for personalized medicine. These algorithms can be pretty complicated and until recently, the huge data volumes and complex data structures prevented us from building such applications.

All publicly traded companies experience the stress at the end of a quarter to close their books and provide a decent outlook into the future. Would it not it be nice to produce a full Profit and Loss (P&L) statement within seconds, discuss necessary accruals and book them, and then rerun the statement, while still in discussion with the auditors? Or to use market trends for energy, raw materials, transportation costs, labor costs, and political situations to calculate the impact on certain results in a simulation, using the full data of an enterprise system, also within seconds?

And the next generation of the Internet is coming, with massive requests for Big Data and speed. More and more devices will be connected and help optimize business processes, but will also demand change in other places. After many years of exploiting the same concept, we have to gain flexibility back before we get stuck. Simplification on all levels is the best preparation for the changes to come.

These scenarios have one thing in common – the underlying applications have to work at a speed which matches our human interaction expectations. This, and the fact that we can invent a nearly unlimited range of new applications – and even business models – should demonstrate the importance of speed in a next-generation database. Speed gives us the opportunity to rethink business practices, forces us to challenge the as-good-as-it-gets attitude wherever possible, and enables us to look further into the future with predictive algorithms and simulations. Speed leads to the removal of latency. To be fast in analyzing historical data is one thing; to be fast in analyzing data while things are happening is something completely different. Think about a new marketing campaign on TV or online, and imagine if we could monitor consumer reactions live. The window of opportunity is different for the various applications, but if we are a little too late, we will miss it.

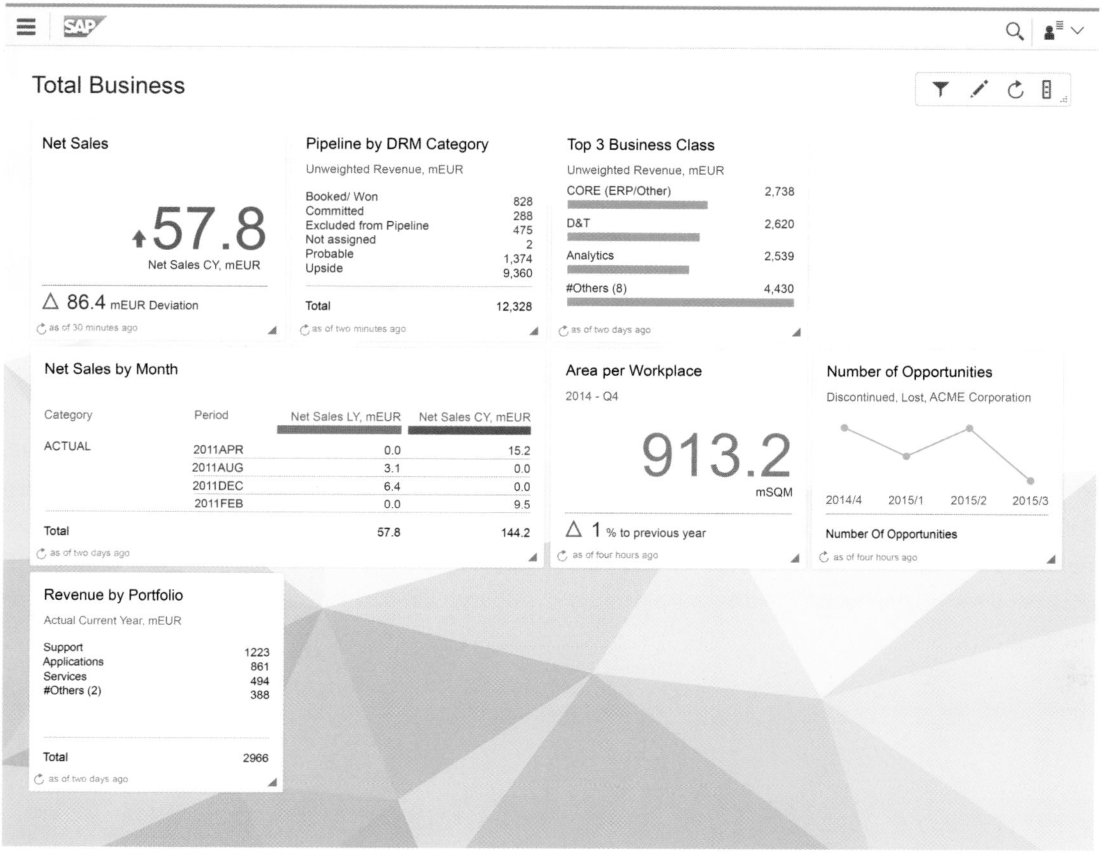

FIGURE 4.1
Personalized dashboard

4.3

The New User Interface

At first glance, there is no direct connection between the database and a User Interface (UI). Yet, remember that with response times close to zero, there is a vast potential for a whole lot of new ideas. For every role in a company, we can define Key Performance Indicators (KPIs) and present them in the form of a dashboard. In Accounts Receivable (AR), the Days Sales Outstanding (DSO) number could be such a KPI. The calculation takes around one second in a large company (>$10 billion in annual revenue) and if the number is being displayed together with a date, the user can decide whether the number is recent enough or should be recalculated. To decrease the DSO number might become a motivation especially when a chart with its development over time is only one click away. In the current accounting system on a traditional database, we would never think about calculating the DSO number on the fly. Now, we can even specify the DSO for each customer class, depending on a particular user profile (Persona). These ideas are applicable to basically all KPIs and help to further personalize the user interaction (see Figure 4.1).

Another example is the P&L statement, which also takes only around three seconds to calculate for a large company. There are now various filter options possible since we access the data on the highest level of granularity – the journal entry line item. It is easy to exclude a certain part of the organization from the P&L statement. All line items, revenue or cost, for this particular business area will be ignored and we get a P&L statement for the rest of the company. It only depends on what we defined in the coding block upfront and not on any intermediate aggregation anymore. For example, in Figure 4.2 the user has chosen three specific products. In the next step, the user could split the series according to these products. All new program designs will provide this type

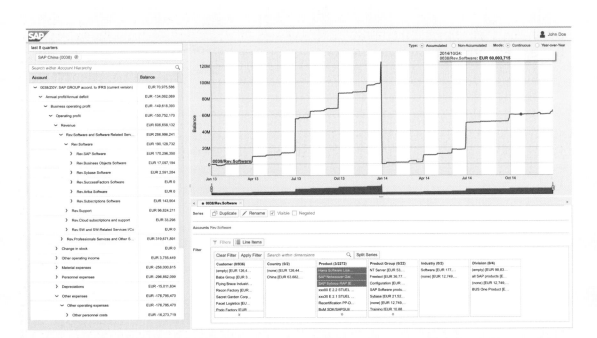

FIGURE 4.2
End users are able to compose their own queries.

of flexibility as a standard feature. In Section 5.4: Sales and Marketing we present other examples of this flexibility.

Simple things, like a chart of accounts, are now available in no time and work as instant coding and filtering assistants. Many workflows have to be revisited, and eventually redesigned, to take advantage of the radically different response time pattern. Reports can also become much more flexible and adaptable for specific situations.

Here, Design Thinking will help tremendously to develop new ways of reporting and customers should share the new ideas (see Chapter 6: Leveraging Design Thinking in Co-Innovation Projects for more details). In Section 8.2: Taking the Speed Challenge, we show how reporting and its corresponding workflows are transformed with the introduction of HANA.

More functionality can be included as long as the navigation and interaction remains clear and

FIGURE 4.3
A flexible search dialog in the Simple Finance application: Manage Customer Line Items

easy to use. With HANA, it became possible to introduce a freestyle search function within the transactions. For instance, we can search for products while creating an offer, find an account number while posting, or look for an interesting data fragment in a transaction table. As an example, Figure 4.3 presents the new and flexible search dialog in the Manage Customer Line Items application.

Furthermore, the user can navigate from business object to business object. For example, this enables the user to find all alternative offers for a purchase or all shipments for a customer without going through the single orders. The user can navigate through all object relationships without using the real transactions and only call the standard transaction to display the object when requested. Section 10.3: Navigation through Enterprise Data explains the workflow of this process.

⇒ **new user navigation techniques are possible with HANA**

Along with SAP Fiori, we can include other visualization applications like SAP Lumira and Geographic Information Systems (GISs) from different vendors in order to build increasingly popular mash-ups which combine different information sources in one appealing UI. One example is shown in Figure 4.4 and another can be found in Section 10.1: Determining the Who, What, and Where of Consumer Behavior.

⇒ **SAP Fiori enables the creation of mash-ups**

Several versions of the same principal functionality can exist in parallel. In order to support productivity, the program functionality has to adapt to different User Interfaces. The simpler the underlying data model is, the easier this development can be mastered. Despite the ongoing trend of users shifting to mobile devices, we still have the back office working with desktop computers with large screens. Many of these users have even multiple screens around their workspace. They can have several applications open at the same time, use a keyboard, and typically possess a high-speed Internet connection. The front-office users often have their main job activities away from the computer (sales clerk, nurse, manager, engineering service person, etc), and for them, the application has to be extremely easy to use and convenient. The mobile applications have to take advantage of the touch screen and work with the more limited display space. Even between tablets and smartphones or smart watches, we have to provide variations in design and interaction (see Figure 4.5).

⇒ **variations of the User Experience (UX) work on workstations, tablets, mobile phones, and smart watches**

The parallel existence of different versions of similar functionality also helps to introduce new versions of an application completely non-disruptively in very short cycles. This is especially helpful for applications offered as a service in the Cloud (SaaS). Therefore, the new SAP Fiori front-end applications constantly go through a comprehensive beta program with experienced customers before they are applied to all customers in the Cloud.

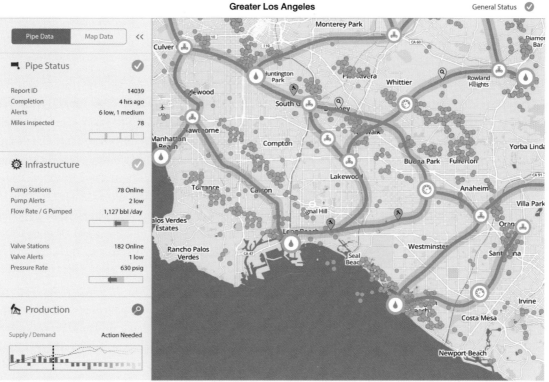

FIGURE 4.4
The FixIt mash-up supports maintenance of hydraulic pumps.

FIGURE 4.5
Fiori applications work on different devices from workstations to smart watches.

4.4

Better Systems with a Lower Total Cost of Ownership

The Total Cost of Ownership (TCO) of a system in an IT landscape is the sum of all direct and indirect expenses associated with the system over its lifetime. The main cost drivers are hardware and software expenses, implementation and deployment efforts, as well as the costs of operation.

How can a database provide a much better platform for enterprise applications, and simultaneously be cheaper with regards to hardware and operational costs? How can a change in database technology result in so many various benefits? One answer lies in the original research idea – what if a database always has zero response times? If we assume this is true, or could be achieved eventually, we would build applications very differently.

A large part of application design and programming concentrates on achieving decent performance. If we can drop these efforts, systems become

significantly simplified. This is the key; with simpler systems we can adapt to external changes, integrate components more easily, connect easier to other applications, reduce maintenance efforts, run much higher volumes on the same hardware, improve mandatory tasks in the data center, and last but not least, build a much better UI. This is what this book is about – showing what is possible with a new database in the world of enterprise applications and its related areas.

The widespread opinion is that for transaction workloads (OLTP), which are presumed write-intensive, we should use a database with row storage, and for analytical workloads (OLAP), which are presumed read-intensive, we should use a database with columnar storage. Every single academic book about databases takes this as a basic assumption. Our research and practical experience show that this is not the case – OLTP and OLAP can easily run on the same database, with better performance for both. The trick is a cascade of application design principles, which in the end allow for even better speed of OLTP than anticipated.

It all starts with an extremely high speed for scanning table attributes. For a full scan of a column, the runtime is proportional to the amount of data to be scanned. The amount of data is equal to the compressed size of the attribute multiplied by the number of rows in the table, and therefore, in the column. The scan speed is, as mentioned before, three megabytes per millisecond per core. This means for an account number (eight characters, compressed two bytes

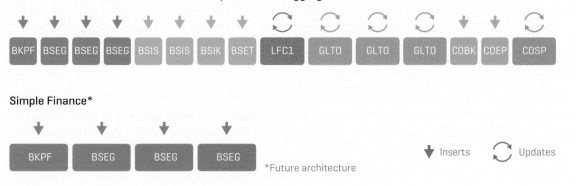

FIGURE 4.6
The simplified data model of Simple Finance (see also Figure 3.4).
Each box in the figure represents a table or materialized view and the corresponding inserts and updates for posting a simple supplier invoice.

per integer), we get a scan speed of 1.5 million values per millisecond when we use only one core, or 15 million values per millisecond if we use ten parallel cores. This is extremely fast. Because the scan speed and the aggregation speed are both so fast, we calculate totals only on demand instead of maintaining them transactionally as materialized views. The on-demand calculation is slightly more expensive than accessing a precalculated total, but happens only if requested and is still sufficiently fast to guarantee that the application remains responsive. The gain in flexibility is discussed in Chapter 8: Highest Flexibility by Finest Granularity. To maintain totals transactionally is relatively costly (direct read and update, sometimes with a database lock). Dropping the transactionally maintained totals (aggregates), we save substantial time when entering transaction data. In the account number context, we maintained two or more totals for each line item. Whenever we maintained totals, we also inserted a redundant line item documenting exactly each aggregation step in addition to the main journal entry.

Because we no longer maintain these totals, we will not have a database locking problem (simultaneous write access for the same table entry) and therefore, can run multiple data entry transactions of the same type fully in parallel, without the need to sequentialize them using a transient data construct.

All of this together reduces the number of database accesses from 15 inserts and updates and 14 reads to 4 inserts for posting the simplest supplier invoice. This is the reason why entering invoices into the database is now 3.5 times faster in the financials system, despite the individual insert in HANA being 1.6 times slower than in a row storage database also in memory. The fact that we can improve the data entry processing speed with a columnar storage database is completely contrary to the aforementioned academic opinion and today's professional experience.

Figure 4.6 shows the removal of all redundant inserts and updates in Simple Finance (without reads). It is very important to know where the advances are coming from in order to trust the achievements in performance. An improvement by a factor of three to four in transaction performance on the database side is phenomenal if we recall the stance of current academic literature and the long fight to arrive at today's performance on conventional databases.

Some of the gains in speed and workload will be traded for better usability in data entry transactions, work lists, select lists, free text search, business object navigation, etc. These significant improvements are only achievable with a database which stores data only once in columnar form and not twice in row storage and additionally in columnar storage. HANA is such a database. It is amazing how much concentration of data is possible by applying simple business rules. As a rule of thumb for compression from an uncompressed ERP dataset to the optimally compressed actual dataset, we saw factors of up to 50, depending on the industry.

In the workload simulator shown in Figure 4.7, using real enterprise data, we can show how the split into actual/historical data and a single replication help optimize the load distribution and balancing. Using the same hardware configuration, we can improve the average user response time, and even the system workload capacity. Like in other web systems with multiple replicas, we can now handle basically any workload demand, and since the actual data footprint is relatively small, we are able to do this in a very cost-effective way.

FIGURE 4.7
The workload simulator highlights performance gains (numbers shown in red and black) when separating all data (left) into actual and historical partitions (right).

⇒ **less overall data footprint (1/10)**

⇒ **very small actual data footprint (up to 1/50)**

⇒ **replication of actual data for workload capacity increases**

⇒ **much simpler data model with less demand for database administration**

⇒ **up to four times higher data entry speed**

⇒ **huge gains in time and space for backup and restore**

⇒ **very high availability through replicas as standby**

The new application architecture was only possible because of the high aggregation speed of the database. This allowed for a domino effect to occur after removing the transactionally maintained aggregates. The TCO for the new Business Suite will be substantially lower than in any previous configuration.

Quantitative Gains turn into Qualitative Benefits

After we acknowledged that all enterprise applications are transactional in nature, we can call them when required, specify our needs, and wait for their response so we can continue with the next step of our work schedule. No question, we have to change our work processes to take advantage of the availability of instant insights, correlations, and forecasts in the window of opportunity. At the speed of thought, we can finally consider simulations (rapid reruns of algorithms with variations in key parameters) or predictive analytics in order to start responsive measures before an event with major consequences even happens. We have moved from the responding enterprise – even in real time – to the predicting enterprise. Many application areas will provide customizable dashboards to give overviews of process queues, exceptions, key factors, and performance indicators. We have to be prepared to spend computing capacity, but we can improve the productivity of our users and the quality of their work enormously. After many years of stagnation or incremental improvements, a new era of transparency, prediction, insights, and creativity begins.

The demand for a real-time enterprise system is decades old. However, with the invention of mobile devices, information has to come in seconds or users will move to other tasks. The instant response became a must, as most interaction, at least information retrieval, will come from such mobile devices. Another reason for a real-time system is based on the fact that more and more sensors are sending transaction data which can be processed instantly, whether we record critical data from engines or we follow the flow of goods by reading RFID tags along the supply chain. Just think about the constant chatter on social networks and its impact on market intelligence.

As we will see, most enterprise application functionalities will become real-time transactions. With this, we have to rethink how we support sales and marketing, how we monitor our products in daily use, and how we gather information about our customers or consumers with regards to our products. We can connect all the information we can access, and create a much simpler IT landscape, with huge improvements for the users of our enterprise systems. The reduction of IT complexity is one of the most common requests in the market. Less complexity does not mean less functionality or less flexibility; to the contrary, a cleaner engineering approach helps to simplify and improve some of the developments from the past two decades.

More timely information lets us take action at an earlier stage at lower cost. Some needs can be covered by standard application offerings, while others have to be developed individually for a company based on the industry characteristics, the location, the size, etc. The faster development speed with HANA-based applications is an additional benefit.

The co-innovation initiative is SAP's answer to this challenge. The idea to complement the standard products with individual application development is not new. However, with the versatility of HANA reaching a new level, the more creativity we can achieve in the partnership, the better the results will be. The Design Thinking methodology of Stanford University has proven to be very helpful in creating an atmosphere for creativity within such joint projects.

Partitioning the Data of Enterprise Transaction Systems by Application Logic

We collect larger and larger amounts of data in today's enterprise systems. Not always do we know how frequently the data will be accessed later, nor do we know exactly the type of access (whether it will be direct or sequential). One approach to optimize the use of storage hierarchies is to monitor all data accesses and distribute the data accordingly. Databases have done this for many years with sophisticated caching, to the extent that the vast majority of direct data access can be satisfied by access to the Dynamic Random-Access Memory (DRAM), and all data is still kept on non-volatile disk or flash storage. The direct access to single items or a group of items is supported by indices which are kept in DRAM. In the past, we used only row-oriented databases for transaction systems, and the concept of caches with aging algorithms was fitting because a large part of the read access was direct and we had a significant amount of updates.

In modern in-memory databases with columnar data storage however, all attributes can be used as an index, and the traditional database indices are not necessary anymore with the exception of the primary key. The high data compression rate allows us to keep much more data in memory than in traditional databases. But again, the question arises whether all data should be treated equally and kept completely in memory in addition to the non-volatile storage on disk or flash. One approach is to analyze data access and try to optimize based on costs whether data should be kept permanently in memory, or only be loaded on demand. But the change in the data access profile, with the dramatic drop in direct access, makes this approach less favorable. Keep in mind how attribute vectors have to be completely in memory, otherwise the processing times would become as unpredictable as before.

Because of the superfast execution of scans, filters, aggregations, and other query types when using an in-memory database with columnar storage, we now build applications in a different manner. These applications will have a contrasting query profile, e.g., fewer inserts, fewer updates, less direct accesses, and many more full column scans and range selects. The optimal use of memory should be based on an application-oriented design of data distribution, rather than a statistical one.

Every application has to define whether its data objects are still necessary to conduct business, and whether they can still be altered or they are still needed to fulfill annual regulatory requirements (e.g., statutory reporting). If none of this is the case, data objects cannot be altered anymore, are not necessary for legal reporting, and most importantly, are not used in the transaction processes – they become historical information. But even this data has to be accessible for regulatory reasons for up to ten years, or for internal statistical purposes for five to

ten years. It is obvious that the two categories of data have totally diverging access patterns. Recall, that these two categories precisely represent the partition into actual and historical data. Following this concept, historical data cannot be altered, meaning that we will not observe further updates. Direct access to an object of the historical category is relatively rare, and most of the access will be of a sequential nature. Not all data objects are equally interesting after they become historical data, e.g., the documentation of stock movements will be of less interest than the analysis of sales orders or Point of Sale (PoS) data. Since all data of an in-memory database is stored on non-volatile storage, and data is only loaded on demand into memory, we can use a purge algorithm to reduce the necessary capacity of DRAM. A possible solution could be that all tables which have not been accessed for a certain amount of time (e.g., a week) will be purged from memory, and only will be reloaded if there is a request. Actually, in the HANA database, this can happen on the column level instead of only the table level. In this case, all the columns which are only necessary to calculate taxes or conduct a workflow (like textual annotations explaining a transaction), will nearly never be accessed again, and therefore, will not show up in memory. However, this only works if the applications are free of any "select *" operations (accessing all columns of a table) and manual nested-loop joins, performed by the application through direct reads.

Once we establish a separate storage partition for the historical part of data objects, we receive some extra benefits. Since per definition no data can be changed, we do not have to archive these partitions on a daily basis. For a typical system, the volume of the historical partitions will be around 4:1, in comparison with the actual partitions. The reduction in backup efforts for the actual data is significant, and high availability of the actual data becomes much easier to achieve. When analyzing data in the historical partitions, we must include data from the actual partitions as well, since certain business objects such as offers, orders, invoices, etc could still be active. This requires the format for historical and actual data to be the same. And, because it is the same, we can apply the caching of partial aggregates or other result sets per partition, and combine them with their equivalents from other partitions.

As an example, we can run a sales data analysis over three years again and again, while data is still coming in through all sales channels, with a response time of milliseconds. The trick is that the filtering and aggregation of the historical partitions and the main storage of the actual partitions (only the delta storage changes during the day) remains the same, and the result sets can be cached for the duration of the analytical session. The same applies to simulations and other mathematical algorithms, including multiple partitions of data. For comparisons of actual data with historical data, we can cache the result sets for the historical data since changes to that partition are no longer allowed, and a P&L statement for the past fiscal year can be calculated once and kept for a year or more. Only when the aggregation logic (hierarchy) changes, do we run a recalculation triggered by the user.

This business rule-based partitioning of data objects is by far the most efficient way for enterprise data. There may well be other forms of partitioning helpful for improving processing times and/or storage optimization. The point is that they have to be defined by the application domain, and not by just monitoring the past statistically.

As a result of this type of partitioning of business objects, we achieve a much smaller dataset for the actual data, which translates nearly linearly to reduced processing times based on the assumption that range selects are by far the most frequent access type. The amount of DRAM we have used in traditional databases for caches is sufficient to hold all actual data in memory. Now, having reduced the DRAM requirement for the actual data, we can concentrate all data on one node on existing hardware, even for the largest companies in the world. In order to balance storage and transaction volume, we can scale up by introducing redundant replicas of the actual data.

The majority of the transactions in modern enterprise systems, where we combine OLTP and OLAP activities, access data in read-only mode. With the short response times for queries accessing large amounts of data, the usage of these read-only applications will increase substantially, yet they can all be executed on one or multiple replicas. Only when, very rarely, an application needs to run on the absolute latest version of the data, does it run on the primary database node. The decision when this is applicable is an application decision, and is not a database consistency question.

Depending on the frequency of database requests for the historical and the actual data, we can assign them to separate database nodes with different hardware properties. As actual data is converted to historical data only very infrequently (e.g., once a year), we only have occasional write activity and so we can use flash memory instead of DRAM.

What does this mean for typical enterprise data objects? All data that remains part of ongoing transactional business belongs to the actual data partition. In order to collect all data of the current year, we start with the data part of the transactional business at the beginning of the fiscal year (open invoices, open customer orders, etc) and add all data which then entered the system until today. All other transaction data can be considered historical. Master data and configuration data always remain actual. We concentrate all business activities, including monthly, quarterly, and yearly reports, on the actual data partition only.

The actual data volume in HANA is typically 2–2.5% of the size of the data volume of a traditional database through the compression of data, no redundant data, fewer indices, and partitioning historical and actual data (this corresponds to a data footprint reduction factor of 40 to 50).

We keep all actual data which was requested at least once over a certain period in memory. Data that was not requested once will be purged from memory. This keeps the actual data volume even lower, yet still provides a nearly 100% availability of data in memory, which guarantees predictable processing times.

Once a fiscal year has been closed, a new partition of historical data will be created and the volume of the actual data will be reduced. Whether the historical partitions remain on one node, or are distributed across multiple nodes, has to be based on the size of the partitions and the type of hardware used.

With this split of transaction data into actual and historical by application logic, we have the chance to organize the underlying hardware accordingly. The vast majority of transactions only access data from the actual partitions. All new data inserts and all remaining data updates will, per definition, only take place in the primary actual partition. Only here will the frequent merge of the delta store into the main store take place. As specified by the partitioning logic, all activities for conducting the business, e.g., take orders, ship, invoice, manufacture, purchase, or close books at period end, rely solely on the actual data. The profile of the data access for the actual partitions has an emphasis on write operations to memory and to the permanent data storage. Only for the actual data do we need transaction logging and periodic creation of snapshots. Log data and snapshots should be on flash or other solid state devices. Since the majority of transactions, data entry, and reporting will be concentrated on the actual partitions, we must provide extra capacity via replication. On a replica of the actual partitions, we can execute any read-only transactions in parallel.

Many applications need to join data from different tables, and the joint execution is faster as long as all data is accessible on a single node through shared memory (Non-Uniform Memory Access (NUMA)). Therefore, we should try to keep all actual data on a single node and in memory. The largest single system with shared memory currently offers up to 24 terabytes of memory. Even if we consider that half of the memory has to be reserved for working memory, all known enterprise systems such as ERP, SCM, SRM, CRM, and PLM are satisfied by this capacity.

The historical data partitions, on the other hand, see mostly range selects or full attribute scans. The data storage does not have to be reorganized for long periods of time (months or years), and therefore, no daily backup is needed. Since the data access is mostly read-only, memory technologies other than DRAM could be used. The historical partitions are created at certain time intervals, and could easily be organized accordingly. The advantage of this is that the user can easily identify which partitions should participate in a report (pruning); only these will be accessed, thus avoiding unnecessary data access. Monitoring the access patterns in a production environment will help find the optimal hardware configuration and data distribution. A simulator using the real enterprise data delivers a first approximation for the best configuration.

From a TCO perspective, it is important to understand that we can use significantly cheaper hardware for the nodes storing historical data, e.g., relatively small blades with less DRAM with no need for high-end flash storage.

The Unification of ERP, SCM, SRM, and PLM for On-Premise and In-Cloud

The most efficient approach to reduce complexity and save costs, while still improving system capacity, is to reintegrate all enterprise systems such as ERP, SCM, SRM, PLM, and Business Warehouse (BW) to some extent, back into one system. The removal of intersystem communications, data synchronizations, monitoring, and eventual corrections reduces the overall workload significantly and simplifies operations. This is a continuation of our idea to remove all the redundancies we have created in the past 20 years for performance or development speed reasons.

This is contrary to the widespread belief that future enterprise systems should be a collection of SaaS components integrated through standard interfaces. The management of redundancy is prohibitive, and extra workload from shipping data back and forth creates a bottleneck. It is fully understood that the components of an integrated system have to be competitive in functionality to the existing stand-alone components. The newfound development speed in the cloud environment allows for this change in product strategy. To a large degree, this depends on the quality of the application platform. With the HANA platform, the reduced complexity in the data model, and the new way of building applications, it becomes only natural to approach reintegration.

This does not mean that SaaS applications outside the scope of the integrated system will be ignored. To the contrary, any meaningful service should become accessible via prebuilt interfaces in order to enrich S/4HANA.

The simplified version of the financial components of the ERP system, Simple Finance, became available in October 2014. Since then, other components of the Business Suite have followed. Simplification meant the full exploitation of the potential of the HANA platform, the reduced data model, no asynchronous update tasks, the quarantining of all unused application components, and most importantly, the introduction of the new SAP Fiori UI.

The business advantages are obvious. The radically simplified new Business Suite allows for the reintegration of SCM, SRM, PLM with ERP. These applications remain fairly independent so that they can still be used as stand-alone systems. But, when integrated, all redundant data structures can be eliminated. The much smaller data footprint enables us to run all applications together on one HANA system. The cost savings in hardware, system separation, and software versions are huge. Managing the single system is much easier.

⇒ **data integration always beats messaging in Total Cost of Ownership**

In the Cloud, we keep all applications permanently updated, and therefore, their separation into independent systems for different development speeds is no longer valid. We have observed that up to 40% of total system workload used to be spent in data exchange between the core ERP system and the satellites. Another important step is the reintegration of transaction reporting, which currently happens to a large extent in the BW.

Based on experience gained from rewriting the application code for the HANA platform and using the HANA libraries and stored procedures, the development speed of new functionality will significantly increase in the future. The Internet of Things (IoT), new business processes for predictive maintenance or multi-channel customer interaction, and the use of mathematical modeling amongst many other innovations will extend the scope of enterprise applications. The update cycles in-cloud will be much shorter than on-premise.

The dramatically simplified data model of all applications and the nearly complete removal of data updates, reduces the risk involved in program updates. Like in other modern web systems, new functionality will be provided as a component in parallel and most new features will be completely read-only. Maintenance has reached a new level in cloud operations. The openness and flexibility of the HANA platform easily allows our partners, startups, and others to build solutions on top of the platform and thus, accelerate innovation.

⇒ **the HANA platform allows for fast development in all application areas**

It has become obvious that the simplified data model has a major impact on the inner complexity of application programs as well. This effect is larger than expected and reduces future maintenance efforts.

In the future, there will be no principal difference between cloud applications and on-premise applications. What runs well in the Cloud could also run on-premise, but not the other way around. It makes sense to offer S/4HANA functionality, after it has proven its superiority in the Cloud, also on-premise. This includes the new maintenance procedures which require more frequent system upgrades than what has become accepted practice in the past. The 18 month cycles and the option to ignore new releases are gone. A special focus on customer developments and new solutions using the HANA Cloud Platform (HCP) for no-touch extensions reduces maintenance efforts further (see Section 4.8: Extensibility).

On-premise customers should consider moving directly to the massively simplified S/4HANA, trust the non-disruptive approach, and start reorganizing major business processes. Customers can also let SAP manage the on-premise system remotely in the SAP Managed Cloud. That all this comes with a huge potential for TCO reduction makes it mandatory to consider a complete renewal of enterprise systems, whether in the SAP Managed Cloud, as a user of the full SaaS offering, or on-premise.

huge potential for TCO reduction

Guided Configuration

A key part of the TCO reduction – especially for the SAP Managed Cloud and SaaS offerings – is the simplified customer onboarding and configuration approach provided by S/4HANA. This Guided Configuration approach reduces the time-to-value and the Total Cost of Implementation (TCI) significantly. The standard implementation package for the professional services SaaS solution on S/4HANA could be realized in less than 50 person days of consulting. Guided configuration can be clustered into five key steps:

1. Cloud trial and discover

In the cloud trial, instant access to a free trial system including interactive guided tours is granted. It has never been easier to experience a solution, the strength of HANA, and the look and feel of the Fiori UI. Only a few clicks away from the initial landing page, users get access to a fully configured, running trial system with sample data. Based on select roles, interactive guided tours are offered showing the benefits of S/4HANA live in the trial system. Users can stop the guided tours at any point in time and explore the solution in more depth on their own.

2. Fit / Gap analysis and configuration

The implementation project does not start from scratch. The starting point is a fully preconfigured, ready-to-run system with sample data. The implementation turns into an agile process including iterative sets of tasks with the goal to fine-tune the preconfigured business processes to the requirements of the customer. Time and cost-intensive blueprinting phases are a thing of the past. Experiencing the preconfigured system performing a gap analysis, together with an agile implementation, mark today's approach to an S/4HANA implementation. SAP Fiori-based, easy-to-use configuration UIs are provided to tailor the preconfigured processes to customer requirements. These UIs follow a business process structure and completely replace the current configuration UI in the Business Suite.

3. Extension and migration

Adding a new subsidiary (a company code) to the system and embedding it in all preconfigured processes is just a click away. Building up a chart of accounts, setting up purchasing organizations with specific parameters such as approval thresholds and tools to perform system extensibility, and migration of master and transaction data is now an easy task.

4. Automated testing

Automated customer tests drastically reduce regression testing efforts in the implementation project. The fully preconfigured system does not only come with preconfigured business processes, but also provides test automata and related automated test scripts. These automated test scripts adapt to the data in the system and to the configuration changes the user has performed via the self-service configuration UIs.

5. Onboarding

Users who are familiar with previous SAP Business Suite solutions will quickly adopt S/4HANA. Training and onboarding material,

online help, and documentation are a part of S/4HANA. This information is provided as an overlay for each SAP Fiori application – right where it is needed.

The key characteristic of the guided configuration approach is its focus on content. Drastic simplification can be achieved because content (such as configuration settings, documentation, help, learning content, etc) is delivered together with the software as one consistent solution package.

Integrated Reporting and Analytics

Being able to fully integrate reporting, analytics, dashboards with key figures, and even statistics into the transactional part of business gives us the opportunity to fundamentally change how we work. Instead of working on data at least a day old, using a data warehouse, or even going outside of the system with a private data mart in order to include statistical functionality like R, we can stay inside the transactional world with no fear of compromising transactional performance.

The read-only capacity is essentially unlimited due to data replication, and parallel processing allows us to stay within the flow of transactional business.

⇒ **increased workload is no longer a reason for a separate information system**

As transaction data is kept on the highest level of granularity, any kind of analysis including associated master data can be performed. From a user point of view, the database appears as fully indexed. Any attribute can be used as an index with good performance, and there is no need for involving a database administrator in order to create attribute indices. The nature of an in-memory database with columnar storage is quite different compared to databases we are familiar with. Users have to develop a fair understanding of the architecture in order to take full advantage of the new possibilities. The sequence of filtering, joining, and aggregating influences the performance significantly, and algorithms should be constructed with that architecture in mind. Thus, the idea to include data scientists in assisting the user departments makes absolute sense.

Let us look at an example of how HANA simplifies a financials application. For many years, we have been trained to follow the hierarchical structures of entity relationship data models, e.g., in dunning we read all customers sequentially, and for each customer, we read the open items and identify the overdue ones in order to calculate the total amount overdue (see Figure 4.8). The number of database calls is at least as large as the number of customers, regardless of whether they have overdue items or not. With an in-memory database based on columnar organization, we can reintroduce classic data processing techniques such as filtering, sorting, merging (joining), aggregating, and calculating with great success. It is amazing to see how much more efficient these techniques can be in an in-memory database. With HANA, we select only the overdue items by scanning the attribute vectors, defining the due date, and for these due line items we select all the attributes needed for the dunning procedure. We then group them by customer, calculate the total amount overdue per customer, and finally read the customer master data. In an example on customer data, we measured the new algorithm as 700 times faster than before, and as a consequence, the dunning procedure becomes a transaction which only runs for a second or two in a large company. Now, we can rethink how we design the dunning process. For example, people in a sales organization can call this transaction from their mobile phone while traveling to their customer and examine the receivables.

⇒ **HANA allows for more efficient non-hierarchical algorithms**

There are several ways to address reporting: ready to use reports of the standard applications, reporting applications with room for customization, individual programs, and generic reporting tools. We are forced to always make the choice between maximum flexibility and time for creation. The SQL and OLAP functionality of the HANA database and the additional mathematics libraries such as the Business Function Library (BFL) and the Predictive Analysis Library (PAL) allow us to use any kind of programming language and presentation layer. The simplicity of the data model and the fully indexed relational tables lead to a greatly reduced programming effort. Another huge benefit is the possibility to integrate unstructured data using the text processing functions.

The applications provide a comprehensive selection of prefabricated reports with some potential for customization. A good example is the new account display transactions in financials. Here, line items from one or multiple accounts can be selected. The accounts could be various contact points with a large international customer or supplier, all customers or suppliers in a geographic location like a city, state, country, industry, or a single account identified via a free text search. The user can easily switch between various layouts depending on the current business task. The extremely short response times should encourage the users to play with data and seek deeper insights.

Specific reporting frameworks, like spreadsheets in accounting and management reporting, are based on domain knowledge and allow for a lot of flexibility using hierarchies, drill downs, graphics, etc; yet, these are limited, by definition, to the scope of the framework. Generic reporting tools provide infrastructure and allow for a lot of flexibility, however, they lack domain knowledge.

All of the above benefit from the processing speed of HANA, and materialized data aggregations like star schemas or cubes are not necessary anymore. Since all reports can reside in the transaction system, we can link them with each other and build cascades of information from dashboards

Old Dunning Algorithm
[Execution Time: 1050 s, Database Calls: ~1,000,000]

```
                    SELECT customers with open items (MASTER);
                    FOR EACH customer with open items:
                        SELECT open account items FROM materialized view (BSID);
                        FOR EACH open account item:
  Loop 1    Loop 2        IF (item is overdue) AND (Sum of open debts > Sum of open credits) THEN
  ~10,000   ~100              INSERT (customer, header, and item) to result table (MHND);
  └─ executions ─┘ RETURN result table
```

New Dunning Algorithm
[Execution Time: 1.5 s, Database Calls: 1]

```
                    SELECT customer, header, item
                        FROM accounting document segment (BSEG)
                            JOIN accounting document header (BKPF)
                        WHERE item is overdue
                        GROUP BY customer
                        HAVING SUM(openDebts) > SUM(openCredits)
```

FIGURE 4.8
Rethinking and accelerating of the simplified and abstracted dunning algorithm in pseudocode. While the old version follows a hierarchical flow from customers to open items to the calculation of dunning conditions in two expensive loops, the new dunning algorithm is based on set processing and the query can be completely executed close to the data.

with key figures, through management reports, down to the listing of line items and their original business transactions, including scanned documents. This was the dream of management information systems 40 years ago and only now will become true. The dramatically improved response times (>100 times faster) will lead to an increase in usage and added value.

4.5

Applications in the Cloud on HANA

L et us continue examining why HANA is an ideal platform for applications in the Cloud. We will go step by step through some of the unique features of HANA making it the best choice for Software as a Service (SaaS). This applies to S/4HANA on the HANA Enterprise Cloud (HEC), the SaaS version of S/4HANA, business network applications, customer and partner applications on the HANA Cloud Platform (HCP), and many new services from startups. This is my [Hasso Plattner] personal view as an academic, and not an official position of SAP. As such, it may not be complete,

and is based on my research experience at the HPI in Potsdam.

Applications offered as a service promise instant availability, easy configuration, elasticity when growing, continuous improvement, connectivity to many other cloud services, a guaranteed service level, and hassle-free operation at lower costs. The service provider has to cope with these challenges by building applications in a proper way, focusing on ease of use and choosing the hardware and the system software carefully – always with the aforementioned goals in mind. The services in the SAP Cloud vary from marketplace or business network applications like ARIBA, Fieldglass, and Concur, to enterprise-specific applications such as S/4HANA, Business One (B1), Business by Design (BYD), and SuccessFactors. The underlying database has to handle large datasets and many smaller ones at low costs with high performance. HANA is an in-memory database with columnar storage. Together with the new developments in hardware allowing for large amounts of memory to be shared by many CPUs with up to 15 cores each (INTEL Haswell), HANA offers some very interesting features which are especially helpful in cloud environments. Most of these we have visited already, but now we will discuss them in the context of a deployment in the Cloud.

The columnar storage architecture allows for the use of massive parallelism. Database transactions can be split into multiple threads running on different cores in parallel. The number of parallel threads varies automatically with the data volume and the current system load. It is easily possible to change the resource provisioning for a given system. Parallelism enables larger analytical or simulation runs even in a multi-user environment.

Dictionary-based compression reduces the data footprint for all types of applications (by a factor of five). All activities such as initial data load, data backup, system restart, or data reorganization benefit from this fact. The disk capacity can be reduced accordingly.

The speed of HANA makes older concepts like transactionally maintained aggregates, data cube roll up, and materialized projections or views superfluous, and thus, accelerates data entry transactions. S/4HANA sees another data footprint reduction by a factor of two because of the removal of redundancies, as well as a significant reduction in size and complexity of code. Other redundant data structures like star schemas are not necessary anymore, and again, reduce the inner complexity of applications and allow for more flexibility. Algorithms are replacing complex data structures and can be changed at any time or exist in different versions in parallel. This opens up completely new ways of designing and maintaining applications, and simplifies the whole data model.

The data footprint reduction by a factor of ten is far more efficient in cost savings than a multi-tenancy concept. Multi-tenancy is still used for relatively small tenant sizes.

By applying dictionary compression, data is converted into integers, which speeds up the database's internal processing. As a consequence, the higher capacities allow cloud applications to service more users for the same hardware costs.

Columns of tables which are not being populated do not take up any space. This feature is important for generic applications serving a variety of industries and locations in one runtime system.

New attributes can be added on the fly without disrupting cloud services. In general, developers on the HANA platform experience faster development and test cycles, which is vital for permanent improvement of cloud services.

All columns of a table can work as an index. The scan speed is tremendously fast, due to compression and integer encoding. The number of database indices is essentially reduced to primary indices, with a few secondary ones. The database management overhead in HANA is much smaller than with traditional databases. Since data is never replicated in row storage and in columnar storage, and operational reporting is conducted on raw data instead of manually created and maintained BW cubes, no further administration has to take place which greatly reduces the workload of the database administrator.

Via federation, called Smart Data Access (SDA), HANA can access data stored in databases outside the system it runs on. Common data like countries, cities, measurements, weather information, historical data, maps, etc can be shared by many applications in multiple systems.

Transaction systems can be split into an actual part (data can change any time) and a historical part (data exchanges only periodically, if at all). HANA uses different strategies for keeping data in memory. For historical data, HANA uses a scale-out approach if the data volume requires multiple physical chassis. For the actual data (ca. 20–25%), HANA uses a scale-up approach via Symmetric Multiprocessing (SMP) systems across multiple chassis if the data volume should require it. The size of the actual data footprint is 40–50 times less than the original on any traditional database (a factor of 10 by dictionary compression and data footprint reduction times a factor of 4–5 by separation in actual and historical).

For extremely high availability, HANA replicates the actual data of enterprise systems and uses the replica(s) for read-only requests. This is the one exception we allow for our otherwise redundancy-free system. Only the actual data of an S/4HANA system will be part of the daily backup routine. Since the historical part cannot change and is only periodically expanded by actual data becoming historical, for longer periods of time no backup is required. This alone contributes to significant operation cost savings.

HANA incorporates text and geospatial capabilities. Completely new applications are possible, and make cloud service offerings, together with HANA's SDA, more attractive.

Sophisticated libraries for business functions (BFL), planning (planning engine), as well as forecasting and simulation (PAL) simplify the application code.

The combination of OLTP and OLAP in one single database on one single image (columnar storage) changes the way we build enterprises. In the case of S/4HANA, this means many reports, planning activities, and optimization algorithms currently running in separate systems can come back and share the transaction tables directly.

HANA is compatible with Oracle, IBM DB2, MS SQL Server, and SAP Adaptive Server Enterprise (ASE), and only stored procedures have to be translated or rewritten. Multi-tenancy via the database for smaller application deployments or generic applications of the business network type supports a lower cost of operation. However, the benefits of multi-tenancy with regards to storage savings and operation costs are less visible in an all in-memory architecture. Virtualization will be used for testing, development, and smaller systems. The effect on large systems is not fully understood yet.

With HANA-based applications following the new principle, the percentage of updates of all database activities is minimal, and the remaining operations are done in insert-only mode – this nearly eliminates the need for database locks completely. All applications modifying tables benefit from this.

The architecture of the data store supports the caching of intermediate results, while new data input is added automatically. This is very helpful for closing procedures or for processes with a rapid sequence of varying queries of the same tables. Result sets from historical data remain unchanged for longer periods of time, and only the data from the actual partitions has to be recalculated. Analyzing PoS, sensor, process, or maintenance data at high speed is possible this way.

Long running batch programs are omitted nearly completely, reducing the amount of management by the user and the service provider. The tasks run in real time as normal transactions.

A separate data warehouse, running only OLAP type applications, is still valuable and well-supported by HANA.

All these features make HANA a very attractive database for cloud-based applications of any type. It is well understood that the user of a cloud-based service does not care about its technical deployment, but will clearly experience the performance and ease of operation. All the early SaaS providers spoke about multi-tenancy as a major attribute of their solution. However, this is only a technical deployment option with cost savings in mind. Hopefully, we have shown that there is much more to supporting SaaS solutions efficiently than just this.

After the transition of SuccessFactors or Ariba applications to the HANA platform, the number of queries has more than doubled, which gives us an indication of how strongly speed determines ease of use. An in-memory database will soon be standard for cloud-based applications. This gives cloud applications using HANA now a huge competitive advantage over databases

storing data both in row and columnar format simultaneously. In order to take advantage of all HANA features, the back-end of existing Business Suite applications had to be partially rewritten, moving parts of the application logic into stored procedures. Since S/4HANA is only based on HANA, this makes perfect sense. A large part of yesterday's application code dealt with transactional aggregation and the creation of redundant materialized views. All this code could be dropped. The majority of the read-only part of the code is new, and received a new UI. This new UI takes the speed of HANA into account wherever possible. One of the breakthrough advantages of HANA lies in the possibility to run OLTP and OLAP type applications in one system. Even in data entry transactions, more and more complex queries play a vital part of the application. However, we must understand that OLTP applications mostly run in single threading mode, while OLAP applications achieve much of their performance improvements via parallelism. Using the same database for both types of applications is one thing, to run them on one database instance is another. In this case, we have to manage the mixed workload. One way is to give the OLTP applications a higher priority, and to restrict the parallelism of the OLAP for times with high system load. Instead, a much better way is to use the replica for most read-only transactions, which are more than 90% of the total workload. Both the primary system and the replica still work on the identical permanent storage system (SSD or disk). Here, we see how important the data footprint reduction is for the actual data.

4.6

The Impact of Organizational Changes

In traditional reporting, the transaction line items in accounting, sales, or logistics are first aggregated by using attributes such as account, country, sales organization, product, date, etc in order to reduce the data volume by orders of magnitude. Then, external hierarchies are applied to develop the final reporting structure. This works fine as long as the structural data does not change. But what if a company changes its internal organization, or acquires another company and wants to include them in the system? As long as we use materialized aggregates, the required data transformation is cumbersome, and it is even impossible to have the old and the new structure simultaneously. However, if the creation of multiple aggregations along different hierarchical structures is purely an algorithm, without any need for materialized data structures, we would have full freedom to create new hierarchies at will and apply them instantly. This is a huge advantage for larger organizations active in Mergers and Acquisitions (M&A) activities or applying organizational changes frequently. The switch from using preaggregated data to sequentially processing the transaction line items seems to be a step back technologically, but is actually one of the greatest achievements today, only possible by fully leveraging in-memory columnar storage. For many years, companies have experienced hardship when changing their corporate structures or reporting lines. The online management information system, with its rigid roll up structures, became a stumbling block, and the approach to produce many variations of an aggregation in parallel using multi-dimensional cubes led to nearly unmanageable complexity when facing organizational changes.

For example, we could always change the structure of the P&L statement to some extent, but not the underlying definition of the G/L aggregation. At implementation time, the rules for the G/L aggregation were fixed, and any change would mean a system downtime for the recreation of a different G/L aggregation scheme in the database. The comparison of the totals over time was also limited – either monthly periods with fields for debit and credit in two currencies each or weekly periods were possible. Since we typically compare the current fiscal year with the previous year, the G/L aggregation structures remained largely unchanged. Our new way is to create the P&L statement and many other standard reports from the bottom up. At the highest level of granularity, the accounting line item, the report structure is fully flexible. Now, we can apply any change at any time not only to the current year, but also to the previous years. Comparisons over time are completely free and can be manipulated dynamically by the user. Many reporting structures can exist in parallel. This is extremely helpful when significant changes in the business occur, and management wants to adjust the reporting accordingly.

4.7

System Considerations

The HPI conducted a series of tests for the best configuration for large and super-large Business Suite systems (as well as for smaller ones) on HANA. We wanted to explore both scale-up and scale-out scenarios with larger SMP systems as well as large blade clusters. We believed that with the current experience in compression, and the options to split data into actual and historical partitions, we could find optimal solutions based on reasonable hardware investments.

As a first observation, we found the dramatic reduction of the data footprint striking. This means we can easily keep the whole actual dataset in memory, and consume less memory than for database caches in the previous system using row databases. For the historical data, alternative storage options are proposed.

The Change of the Database Usage Pattern

In Figure 4.9, we compare five database usage patterns. This figure shows relative numbers for the different access types. On the left side is the TPC-C benchmark, which is used in many database performance tests. It shows a significant portion of updates. In the middle, we have the pattern of a typical SAP ERP system on a traditional database (separated by OLTP and OLAP), with the major difference being the smaller portion of updates [KKG+11]. The high percentage of direct reads is characteristic for data access via indices like match codes, secondary indices, and application-controlled joins. On the right side, the pattern for an SAP Business Suite system (separated into classic and simplified) optimized for HANA is quite different. The smaller percentage of inserts is based on the removal of redundant transaction data and the percentage of updates is now very low – totally different in comparison to the old TPC-C benchmark. The percentage of direct reads also decreased because all updating of aggregates was removed, and joins done by the application were eliminated. On the other hand, the absolute number and the percentage of range selects or full table scans increased. Therefore, it does not make any sense to measure the HANA database performance for enterprise applications using the TPC-C benchmark. Some database literature has to be rewritten, especially where OLTP systems are characterized as write-intensive. Modern enterprise applications are mainly focused on read access, with a tendency towards more sequential data access.

When we look at the distribution of database requests in a simplified ERP system, in comparison to an ERP system on any traditional database, we find the following:

› database updates are nearly eliminated
› the number of database inserts is reduced by a factor of 5
› the number of database range selects is increased by a factor of 2–4
› the number of database single reads is reduced by a factor of 5

This is a completely different database workload pattern, with a significant impact on the system setup. Columnar storage is extremely fast for scanning and aggregation of large tables, and more importantly, runs these operations massively in parallel. This parallelism gives us a new dimension to consider in a multi-tenant setup.

Once we reintegrate the reporting applications, which were moved to a data warehouse system only for performance reasons, this trend will continue. Modern in-memory databases with columnar storage serve both OLTP and OLAP systems equally well. The much more

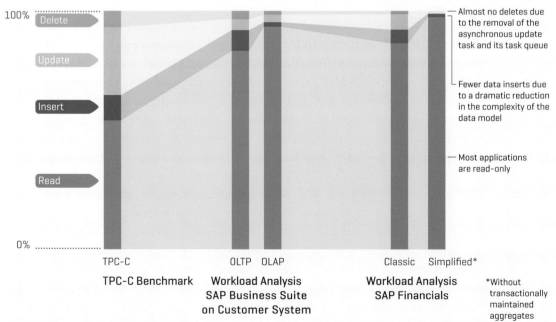

FIGURE 4.9
Changes in database usage patterns

important distinction for the database operation will be whether data can be changed, is accessed directly, is read-only, or is mainly accessed sequentially. These characteristics determine the storage requirements, need for parallelism, aging concepts, cache-coherence considerations, etc. The new enterprise applications on HANA will create a new benchmark database usage profile.

Modern in-memory databases with columnar storage serve both OLTP and OLAP systems equally well.

System Landscape for S/4HANA

A modern server system consists of one or several chassis, which are connected with an internal network. The term "chassis" refers to a single board with two to eight sockets. Each socket has one CPU with 2 to 15 cores (4 to 30 threads) and is surrounded by DIMMs for DRAM. The term "blade" is used for an independent chassis working as an individual server. In an SMP configuration, any core of any CPU can access all memory of all CPUs even across different chassis, while cache coherency is maintained by the operating system and the hardware. In a blade cluster, several independent servers are logically controlled by the database. A node is a logical unit and may refer to a CPU or group of CPUs.

The access to data in memory is not totally homogeneous. For reference, we use the following latencies: 100 nanoseconds for local data access, with this being for all cores of a given CPU, to data belonging to that CPU. With this latency, a

core can access up to 768 gigabytes of DRAM. 200 nanoseconds for access to data belonging to a first neighbor CPU on the same chassis, and 210 nanoseconds to the other CPUs on the same chassis. Data access to memory residing on another chassis within the same system takes up to 3,000 nanoseconds.

Some SMP systems provide a much faster connection between CPUs on different chassis, bringing down latency to 500 nanoseconds or less, by using an extension of the NUMA architecture or a fully proprietary network. With such a system, data access time varies between 100 and 500 nanoseconds according to relative memory location.

We evaluated the options using an SMP system together with a cluster of blades. Let us assume an SMP system has eight chassis, with four CPUs each, and 15 cores per CPU. The total amount of cache-coherent shared memory totals up to 24 terabytes. The total number of cores is 480. The memory could be doubled with the use of 64 gigabyte DIMMs.

For our case study, we used an anonymized version of one of the largest SAP ERP systems in the world, with 100 terabytes on a traditional, row-oriented database on disk. Converting the database to HANA reduces the data footprint through compression by a factor of five, and through the redundancy-free data model by another factor of two. The data is only stored once, mostly in columnar storage (with some exceptions in row storage), but never in both, as found in all other databases. The remaining data has the following properties:

> only approximately 20% of the data is needed to conduct actual business, 80% of the data is only of a historical nature, and cannot be changed anymore – therefore it is definitely read-only. Companies can keep ten years of data in the ERP system online. We are now down to 1/50 of the original data footprint for the actual data.

> all other data can be considered historical and as historical data cannot be changed, we must organize this data differently, e.g., no locking, no delta store, no secondary indices, and theoretically, no direct access. Every year there is a new historical partition.

To produce a P&L statement for the current year, we need to access the actual data partition. For the past year's P&L statement, we only have to access the historical data that corresponds to the last year because P&L line items are strictly annual, and we can cache the result sets because changes are not possible. Most transaction applications benefit from this radical reduction of the memory footprint. The same applies to backup and recovery operations. It is therefore very important to work

on the application-specific separations into actual and historical data. Here are a few examples of programs only in need of actual data access (master data is always actual):

> P&L
> balance sheet
> open items
> dunning
> shipping
> invoicing
> project current year

And in need of both actual and historical data access:

> project history
> revenue development over five years

We can analyze the data access profile for each table and partition of a business object in order to find an optimum distribution of actual and historical data, according to their access patterns. A goal for the optimization could be to keep the actual partitions of all tables for one business object in the memory of a single CPU. This guarantees that the most frequent join between the transaction line items and their business object header takes place on one chassis. Since we know the other typical join partners of a table, we also want to keep them together in the memory of a single CPU, or at least on one chassis. This requires the HANA database to interoperate with the operating system in order to pin data and threads to the right locations.

The primary activity in the database is, as we learned earlier, scanning attribute columns in order to identify tuples to work on. The most critical operation is a join between multiple tables. These two operations, including aggregation, should run with as much parallelism as possible. The data shipment from chassis to chassis in an SMP configuration is faster than between independent blades. Nevertheless, specialized effort is needed to optimize joins across chassis or blades. This has to be supported by the HANA optimizer.

Now, we can easily see that only the actual data has to be stored on a larger blade (e.g., four sockets, six terabytes) or on an SMP server, incorporating multiple chassis if more storage capacity is required. The reduction of updates minimizes the effort to establish cache coherence across multiple chassis. Cache coherence is necessary in a multi-CPU configuration to keep all caches (level 1, level 2, level 3) of all CPUs synced after inserts or updates occur. However, cache coherence is only required for the server hosting the actual data partitions.

Should the workload require an even higher number of cores, we can replicate the system for actual data and run read-only transactions on the replica, which basically doubles the capacity since most of the transactions are read-only.

Currently, there is no known ERP system in need of more than twelve terabytes of DRAM for the actual data. The aforementioned SMP system with 24 terabytes of DRAM in total is sufficient for this kind of actual data footprint and provides ample space for in-memory processing. The advantage of using a multi-chassis SMP system for actual data, versus splitting the data by application area, lies in

the reduction of complexity and the use of a faster internal network.

On the other hand, the historical partitions could reside on smaller blades (less DRAM) in a scale-out cluster, and could be managed by HANA. Since we do not have any inserts or updates to deal with, the shared scale-out approach works just fine. For the historical data only attributes which are used for reporting are kept in memory. How much of the lesser used data should be purged depends on the individual needs of a company. For higher processing speed we could repurpose unused chassis of the SMP server as stand-alone blades and experience faster data transport between these blades. For permanent storage of actual data we should use SSDs, while for the historical data, normal disks are acceptable. The combination of scale-up (actual data) and scale-out (historical data) approaches will provide a good setup. Together with the intelligent distribution of data across blades (and perhaps inside blades as mentioned above) a very cost-efficient and easy-to-manage system landscape is possible.

The combination of scale-up (actual data) and scale-out (historical data) approaches provides an effective setup.

4.8
Extensibility

Enterprises want to create value and achieve competitive advantages through optimized business processes. To achieve these goals, they have to adapt their enterprise software according to their business needs. Despite many years of experience with SAP enterprise solutions, we have not yet met a single customer without extensions. The need for faster cycles of innovation drives cloud-based extensions, e.g., smart metering in the utilities industry. Moreover, custom extensions must be loosely coupled with respect to core business processes: on the one hand, they need tight data and process integration, while on the other, the software lifecycle of extensions must be independent from that of S/4HANA.

In general, extensibility covers a broad spectrum of topics which allow customers and partners to adapt standard business software to their business needs. This spectrum spans topics such as business configurations, layout adaptations of UI forms and reports, custom fields and logic, custom terminology and translations, and customer-specific applications. In cloud products, extensibility has to follow two paradigms:
> the target Persona is the business user
> extensibility must be end-to-end and UI-driven

The business user (also known as the key user or power user) is a business expert who works in a Line of Business (LoB) with the responsibility to adapt software to the needs of business departments. As this person typically has no in-depth IT technical knowledge, all extension steps need to be covered by tools which hide their complexity and technical details.

Extensibility tasks in S/4HANA are UI-driven and can be triggered in a context-aware fashion from the application UI. The execution of extensibility tasks is performed by the business users via the extensibility tools. These tools allow the users to add new fields and adapt application terminology. To hide their technical complexity, these tools work end-to-end, covering all technical steps from database schema modification up to UI adaptation and include all software layers in between.

In general, there are two possibilities for where an extension of an S/4HANA application can be located: either in the same system as the corresponding application (in-app extensibility), or in a separate extension platform (side-by-side extensibility). Both possibilities are valid and have their use cases as we will detail in the following.

The extension process is designed in such a way that there is a clear separation between the in-app

The need for faster cycles of innovation drives cloud-based extensions.

and side-by-side extensibility choices. As a result, the business user is confronted with only one of these options. From the end-user perspective, however, the difference between the in-app and the side-by-side extensibility will be hidden, resulting in a uniform extensibility experience.

In-App Extensibility

The following use cases for S/4HANA extensibility are realized with in-app extensibility:

Field
Customer-specific fields which are part of SAP tables are added to a business context of an application and its UI.

Table
Customer-specific fields which are not part of SAP tables are added to a business context of an application and its UI, i.e., customer-specific database tables are created.

Business logic
The behavior of applications and processes is enhanced.

Report
Analytical reports (maintenance of data sources, definition of business queries, and personalization of reports) are adapted.

Form and email
Print forms and email templates are adapted.

Extensions are created and tested before the extension is active (live) for all users in the production environment. S/4HANA in-app extensibility supports this workflow by providing dedicated sandbox environments combined with an explicit transport-to-production system capability.

The first benefit of in-app extensibility is better performance (lower latency, no additional data transfer) as compared to a side-by-side scenario. Moreover, it allows for direct access to HANA features and libraries. Finally, in-app extensibility allows for integration with application engines such as the HANA rule engine, enabling extensions to access the full business scope of these applications.

Side-by-side Extensibility

In-app extensibility allows for the addition of fields, tables, and business logic to an already rich and powerful application object model. For more complex scenarios, a Platform as a Service (PaaS) approach for extensions is necessary as it allows for the creation of scalable extension applications with a low TCO and a separate software lifecycle.

Extensibility via the HANA Cloud Platform (HCP) is intended for business users and developers, and offers a high degree of flexibility. The HCP is an in-memory PaaS offering from SAP. It enables

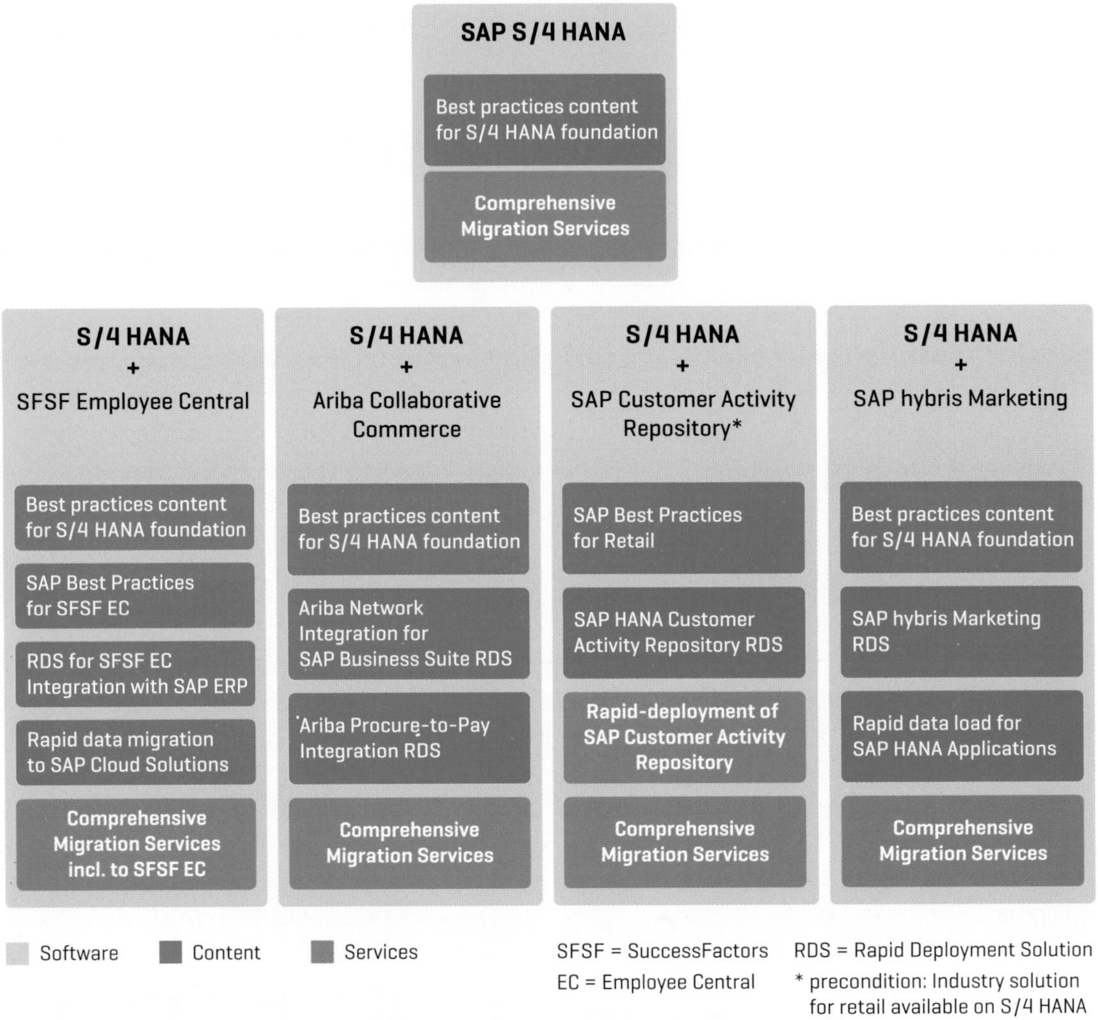

SAP S/4 HANA

Best practices content for S/4 HANA foundation

Comprehensive Migration Services

S/4 HANA + SFSF Employee Central	**S/4 HANA** + Ariba Collaborative Commerce	**S/4 HANA** + SAP Customer Activity Repository*	**S/4 HANA** + SAP hybris Marketing
Best practices content for S/4 HANA foundation	Best practices content for S/4 HANA foundation	SAP Best Practices for Retail	Best practices content for S/4 HANA foundation
SAP Best Practices for SFSF EC	Ariba Network Integration for SAP Business Suite RDS	SAP HANA Customer Activity Repository RDS	SAP hybris Marketing RDS
RDS for SFSF EC Integration with SAP ERP	Ariba Procure-to-Pay Integration RDS	**Rapid-deployment of SAP Customer Activity Repository**	Rapid data load for SAP HANA Applications
Rapid data migration to SAP Cloud Solutions			
Comprehensive Migration Services incl. to SFSF EC	**Comprehensive Migration Services**	**Comprehensive Migration Services**	**Comprehensive Migration Services**

■ Software   ■ Content   ■ Services

SFSF = SuccessFactors   RDS = Rapid Deployment Solution
EC = Employee Central   * precondition: Industry solution for retail available on S/4 HANA

FIGURE 4.10
Overview of packages based on S/4HANA

customers, partners, and developers to build, extend, and run applications on HANA in the Cloud. It can be used for extensions of cloud and on-premise applications.

There are two main use cases in S/4HANA where the HCP is used as the extension platform:

New and enriched UIs

Customers and partners often want to create completely new UIs on top of the S/4HANA open APIs. Another important use case is customized or partner-specific UI theming with the use of the UI Theme Designer. The HCP provides the relevant environment and tools for these use cases.

New applications

In most cases, new business processes and scenarios cannot be addressed with in-app extensibility. Side-by-side extension applications are a means to allow the highest possible extension flexibility. Extension applications on the HCP consist of static resources and a connection to a back-end system using on-premise or on-demand web services (REST). SAP Fiori applications with a data connection to an OData data access service exposed by the S/4HANA back-end system in the Cloud are an example of applications which require side-by-side extensions.

Building extensions in a side-by-side scenario is easy. S/4HANA integrates out of the box with the HCP via a preconfigured SAP Cloud Connector. Moreover, SAP Gateway, which exposes OData data access services to the HCP, is already installed and preconfigured in S/4HANA.

4.9
Solution Packages

To satisfy specific customer needs S/4HANA is offered in the form of solution packages. These packages represent holistic, end-to-end solutions that address specific industries and the needs of Lines of Business (LoBs) with clear and quantifiable value propositions. Solution packages allow S/4HANA to be extended with further SAP solutions, SAP SaaS cloud offerings, and industry-specific applications.

S/4HANA solution packages consist of three major bundle categories (see Figure 4.10):

1. Software

S/4HANA and accompanying SAP solutions, such as SuccessFactors Employee Central or Ariba Collaborative Commerce

2. Content

S/4HANA Foundation Rapid Deployment Solution (RDS) and additional solutions, such as SAP Best Practices, RDS for SuccessFactors Employee Central Integration, and Ariba Procure-to-Pay Integration RDS

3. Services

comprehensive migration services

SAP helps customers to accelerate and simplify their implementation of S/4HANA and related bundles with preconfigured content based on RDSs and a range of right-sized services. RDSs are end-to-end modular solutions that are affordable, fast to implement, and ready for consumption. RDSs are available for deployment on-premise as well as in the Cloud.

New customers can benefit from S/4HANA software by implementing content with the S/4HANA Foundation RDS. It provides a baseline for running business processes and is based on SAP Best Practices, which consist of tested configuration and implementation content, methodologies, and step-by-step guides, thus minimizing the effort for running processes during installation. S/4HANA Foundation RDS also provides data migration services for migrating legacy data.

Existing Business Suite customers can use the rapid database migration of SAP Business Suite to HANA for a quick and seamless migration of their existing SAP Business Suite installation to S/4HANA. The rapid database migration of S/4HANA is an end-to-end solution that leverages out-of-the-box accelerators with a predefined scope. Moreover, it provides a standardized approach and automation that eliminates guesswork and helps with on-time delivery while meeting of all security requirements for easy migration to HANA.

Both new and existing customers can further leverage additional RDSs like the SuccessFactors Employee Central Integration to the SAP ERP RDS, or the Ariba Network Integration for SAP Business Suite RDS. These additional RDSs allow existing and new customers to extend and integrate business processes into the SaaS cloud applications SuccessFactors and Ariba in order to realize end-to-end solutions for LoBs and industry needs.

Helping S/4HANA customers in retail to merge relevant customer and inventory data that previously were spread over multiple applications, the SAP HANA Customer Activity Repository RDS delivers pre-built analytics to provide real-time visibility of metrics such as sales, inventory and multi-channel effectiveness. It provides a unified view of the customer, helping to drive consistent experiences across all channels and improve business efficiencies, profitability and brand perception.

Sifting through market data can be time and resource intensive. Bringing together data from financials, social networks, and websites, SAP hybris Marketing RDS leverages SAP HANA to help businesses understand their customers' preferences and needs and identify the right target audience for their marketing and sales activities. Moreover, it allows businesses to run real-time segmentations for large customers, enabling them to simply and efficiently manage targeted marketing or a sales campaign.

CHAPTER FIVE

REBUILD AND RETHINK

All that we have looked at so far had already been in existence or has now finally become affordable. Yes, the new delivery model in the Cloud is enticing, the speed of transactions is mind-boggling, real-time analytics are possible, and enterprise applications are vastly improved – yet, what are the real breakthroughs on the horizon? In order to identify potential innovations, we must re-examine business processes using the Design Thinking methods that we will detail in Chapter 6: Leveraging Design Thinking in Co-Innovation Projects. The groundwork of data collection is mostly complete and we can build upon this data. Instead of moving data between systems, we learned that we should instead integrate the new enterprise applications with other cloud offerings, and build a loosely bound federation with one standard system – the HANA platform. SAP has begun many co-innovation projects, in which the customer contributes the domain knowledge and SAP provides the technical expertise. The advantage of the Design Thinking methodology is that a diverse group of people is able to work together to identify unsatisfied user needs, collect facts from a wider circle, develop ideas, build prototypes together, and start to iterate. It is important that this heterogeneous group stays together, as all participants will learn during the design process and become able to contribute more and more efficiently. The classic approach of analysis, development, documentation, testing, and production in a waterfall model has been outdated for years. However, we still observe a strong separation between the worlds of business professionals and software developers nowadays. It is understood that time is an issue and that commitment to a project is not easy to manage, but the more a heterogeneous team is able to work together, the better and sooner results are produced.

⇒ **the Design Thinking methodology helps to innovate**

⇒ **business people and software developers work in a team on live data**

This book should give you an idea of what is technically possible, what others have already achieved, and the role that the HANA technology can play, either as the foundation of SAP's enterprise systems or in completely independent application areas.

Some underlying concepts of HANA are greater than 20 years old, yet, only now, have they become feasible due to new hardware technology. The same is also true for business ideas – many concepts could be thought of, yet not realized in the past. Perhaps now is the time to realize these ideas. The HANA platform offers an unprecedented wealth of new, easy to use functions in one system, allowing for rapid prototypes that can become the starting point for developing new applications and business processes. New applications will be more intelligent as a whole. They will interpret facts instead of merely reporting them, predict the near future, show alternative possible actions, and combine data from different sources such as text and structured data.

⇒ **new applications will be more intelligent**

⇒ **instead of documenting the past, predict the future**

In this chapter, we present a brief overview of the most significant opportunities that come with the HANA platform. We begin this chapter with the vision and first steps of the boardroom of the future and then dive into more specialized topics in later sections.

5.1

The Boardroom of the Future

Under the assumption that we can run nearly any report or analytical application in real time (under the eight seconds of maximal human attention span), the notion of running board or other executive meetings based on data in static presentations is rendered obsolete. With little need for assistance from an expert user, all relevant information can be at a manager's fingertips, with further drill downs on request, triggered by the ongoing discussion. Standard key figures and high-level information come as briefing books in the form of a dashboard.

The major advantage of this is that any presentation will happen in real time on live data, and discussion by executives will permanently sharpen in focus, and thus, improve the knowledge base. Hardware for large screens is constantly becoming more affordable, and the ability to play with the data will educate all participants. Large displays with touch control and the ability to draw on the screen will establish an interactive atmosphere. Notes from the meeting will be instantly recorded by the communication system to trigger further actions. The boardroom of the future harmonizes with Bill Gates's vision of "information at your fingertips" [GMR96].

Imagine yourself in the position of a leader of a large enterprise. You have to keep up with day-to-day news from operations around the world, daily sales figures, financial and commodity markets impacting your business, and all the people who want to meet with you. In addition, you and your team are also tasked to manage the strategic direction of the company.

Here is how Strategic Business Management works in most companies today. The operating units are called into the corporate office at regular intervals for business reviews. Business unit leads and their staff spend countless hours pulling reports from various sources, compiling spreadsheets, identifying issues and their root cause, highlighting opportunities, and finally assembling a set of key insights into a pre-read document and a presentation to be shown in the meeting.

The over-scheduled, jet-lagged executive team is then presented with a large presentation of facts, and challenged to connect dots, or identify issues and opportunities outside the curated content that is presented by the business units. Figure 5.1 summarizes the different points of view in a boardroom meeting.

CFO/CEO/COO
Leaders who want to harness the power of their team with the right data at the right time need timely high-quality insights that will enable questions, discussions, and decisions in real time.

Division/Hub President
A leader who is responsible for the financial, brand, and talent health of a division needs a way to react and predict changing business conditions as they occur.

Division/Hub Staff
An astute connector who is good at getting things done is eager to provide business leaders with an in-depth understanding of local business conditions.

Global Functions
Experts need accessibility to data in order to globally balance short-term and long-term strategic issues.

FIGURE 5.1
Points of view in the boardroom of the future

If and when there is an opportunity for a question or a request to see the information presented in another way, in more detail, or enriched with related facts, the presenter has to rely on his or her staff to find the answers. For example, "Show me the net sales for all countries except Venezuela and Argentina," or, "Can we see a trend for a product category for the last four years?" By the time the information is available, new issues have arisen and business has moved on. As one executive we interviewed put it, "We ask a question around the data, and it takes six weeks to get an answer."

The challenges of running a real-time business become even more challenging when it comes to making strategic decisions for the future, such as "What happens if the oil price continues to fall?" or, "What if we increased the price of a product category?" It feels like we are trying to drive a car by looking into the rear-view mirror.

So, what did we find senior leaders needed? Most importantly, they want to discuss the state of their business and make instant decisions (see Figure 5.2). From a technology perspective, this translates into:

> on-demand drill-down – "Have the answer at your fingertips to explain an issue."
> monitor and alert – "Connect the dots and highlight what I need to know about so we can fix it."
> anticipate what might happen – "If the oil price continues to drop, what is the impact on my business?"
> predict to plan – "If I can see the potential cycle, I can decide on the next move."
> evaluate alternatives – "Raise prices, cut overhead,

or increase advertising? What will help me best meet my business goals?"

You might say, "Just give them a dashboard." When the senior leaders drew up their dream solutions, we found that their human and business needs go beyond a dashboard. Senior leaders do not want to dig – they want their staff to analyze business and talk about Key Performance Indicators (KPIs) in a consistent format in their meetings. The business staff want to come prepared to the meeting, in order to not squander valuable time digging for data.

The data used for business reviews often comes from many internal and external sources, for example, macroeconomic data might come from Bloomberg, finance data might come from SAP's Business Suite, marketing data might come from Nielsen, and so on. Out-of-date data often presents another challenge, and data volumes make real-time reporting impossible. This translates into additional business needs:

> a method to design the data model and sourcing in a manner that meets business user needs
> a consistent format for presenting KPIs – for example, by product, geography or key customers, or across all countries
> capabilities that allow preparing and presenting in the meeting but also for interactive drill downs
> a way to capture key issues for follow-up after a decision was made

How will business review of the future look like? Instead of lengthy preparations, ad-hoc business reviews are highly desired, in which with minimal human effort, a standard set of KPIs are

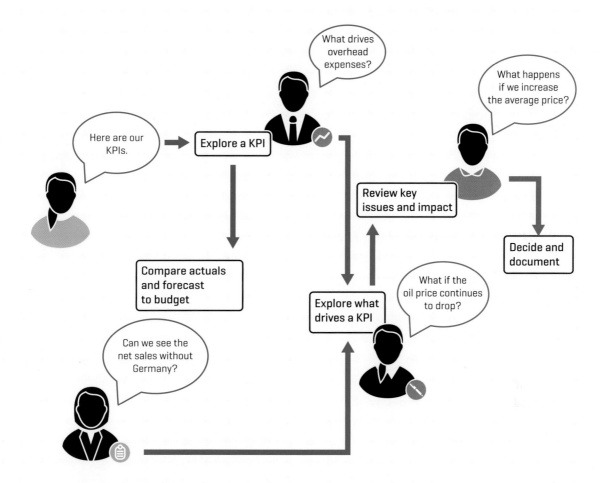

FIGURE 5.2
A typical boardroom discussion

assembled and serve as the common starting point for issues and opportunity-based discussions.

The Business Review of the Future

SAP's solution to reinvent business reviews of the future has two major components:

> a customizable interactive dashboard tailored to the requirements of the executive managers
> a bookmark gallery for storing and organizing relevant information and charts

This solution also offers a communication channel for publishing the documents to be pre-read, gathering feedback, and distributing relevant meeting information.

The Interactive Dashboard

Business reviews of the future are no longer based on static charts and visualizations. Our solution provides customizable and interactive dashboards for various executive roles. Figure 5.3 shows an example of entry dashboards for Chief Financial Officers (CFOs). This finance dashboard allows for easy comparison between Sales Growth, Profit & Loss (P&L), and Balance Sheet items. For example, Price influences Sales Volume but also has a direct impact on P&L, thus, the Gross Profit as the percentage of Sales (Gross Profit % Sales) is a good complementary indicator.

The interactive dashboard represents a source of information for efficient reviews of business indicators. Executive managers can get an overview of their latest company performance on a daily basis which supports the identification of potential issues as soon as they arise. Presenters can pull their dashboard contents into their slides. The dashboard can also be used during meetings to investigate problems or answer additional questions in real time, e.g., by drilling down into details along all dimensions. For instance, an unexpected drop in revenues can be traced by drilling down into regions, countries, business units, and product lines.

In general, KPIs depend on business drivers, and the dependencies of a certain KPI are individual for each company. Furthermore, these dependencies might change slightly over time. However, it is possible to define a map of drivers relevant to KPIs with a Value Driver Tree. Analyzing changes to driver status can be used to estimate changes to KPI values. Figure 5.4 shows such a Value Driver Tree which illustrates how Net Sales is driven by Volume, Gross-to-Net (indirectly via Pricing), and Foreign Exchange. In turn, Net Sales is an indirect value driver for the Operating Contribution as the percentage of Net Sales (Operating Contribution % Net Sales), mitigated by Margin as the percentage of Sales (Margin % Sales).

Business reviews of the future will no longer be based on static charts and visualizations.

FIGURE 5.3
An interactive, customizable dashboard view for Chief Financial Officers (CFOs)

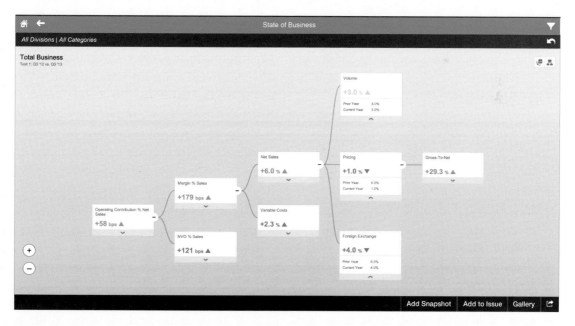

FIGURE 5.4
A Value Driver Tree for an extract of a Profit & Loss (P&L) statement

The Bookmark Gallery

Dashboard views can be used for various purposes: storage for potential future discussions, meeting preparations, reports of issues and the success of mitigating measures, etc. The bookmark gallery enables users to bookmark and organize charts and information of interest for future usage. Bookmarking a dashboard view is not just storing a link. A dashboard bookmark contains a snapshot of the charts and all related information. This allows users to compare the state of a KPI from the creation date with its state today and visualize the differences. The user can then investigate the effectiveness of measures and their potential effects on various KPIs. The bookmark gallery offers a structure that allows for the creation of folders for any upcoming event. Standard chapters for reviews can be defined as well as individual chapters containing snapshots showing issues or decision needs. Comments can be added and read by authorized users to achieve agreement on mitigating measures.

Enabling Interactive Business Reviews

Renewing the process of running business review meetings with the goal of faster decision making requires accurate, real-time interactive data, sufficient flexibility for investigation, reliable prediction of future enterprise performance, and collaboration tools tailored for each specific company culture. HANA is the basis for the realization of these requirements through dashboard views.

Real-time

Having instant access to all line items that affect a KPI enables calculations of the required aggregations on-demand. As a result, dashboards that show all key figures based on up-to-date data can be created. In case of a business issue, executives, business experts, or staff members can identify any related information and take the required actions immediately.

Data integration

The required information for the KPIs can be fetched on demand from various data sources, including all HANA-based SAP offerings such as HANA-based Business Warehouse (BW) systems and Business Planning and Consolidation (BPC) software.

Dynamic and interactive visualization

KPIs are presented visually when applicable. The full use of graphical tools, such as SAP Lumira or other visualization assets, makes data exploration more appealing and efficient. Especially powerful interactive visualizations are an important aspect, and are enabled by pushing down complex computations to the database layer so that expensive data transportation is no longer necessary.

Powerful forecasting

HANA provides a set of libraries that support business functions and advanced predictions. SAP's solution for the business review of the future utilizes the Predictive Analysis Library (PAL) to compute predictions. For instance, performing what-if analyses incorporates the prediction of several measurements so that the best alternative can be selected. Exploiting the power of HANA, such predictions can be performed quickly within review meetings.

Information at Your Fingertips

Many applications can participate in the boardroom of the future, but they must adhere to certain performance standards. Any regression to past response times of minutes or greater, would make participants nervous or bored, a bad environment for innovation. Angry or bored participants will simply choose to not attend the next time, and it is hard enough to set up meetings across different areas of responsibility in a larger organization.

Thus, the time spent on preparation for board-level presentations, and the inclusion of information from the business systems therein, is significant in most companies. This time can be better spent on building applications for interactive discussion based on transaction data, instead of passively using spreadsheets or presentation software on private data marts. This style of supporting the discussion within management allows second and third questions to drill deeper into the facts of a topic.

Once managers are familiar with this style of meetings, the quality of the standard display of relevant company information will also automatically improve step by step. The ability to drill down to the highest level of granularity is necessary to support any findings on aggregated levels. The same applies for regular meetings with a board of directors or supervisors. In Section 8.2: Taking the Speed Challenge, we showcase HANA's ability to beat the eight second rule for attention span even for large datasets and complex queries.

⇒ **finally, information is at your fingertips in management meetings**

Playing with business data interactively will improve awareness of facts and developments, assist reactions to changes, and enhance projections into the future. If we can reduce the amount of presentations, and replace them with easy to understand analytics and prognoses for the future, all participants will acquire a better understanding of the information. This results in better decision making. The experts will be able to develop assumptions and validate them in similar scenarios, or articulate open questions to be discussed in the next meeting. The ability to ask second and third questions – with immediate answers – will lead to a dialogue between the experts and decision makers. Even the application systems can contribute with predictions and proposals for action to spark new ideas.

It was eye-opening to visit a large retailer which uses HANA for Point of Sale (PoS) data analysis. To see retail experts in a boardroom of the future setup, discussing product performance while operating 70" touch screens set up in a semicircle, navigating through various graphical representations of data, and using electronic pens to mark interesting signals while considering the on-screen information is a testimony to the validity of the concept. The participants confirmed the superior results of the discussions, and expect a rapid development of such presentation forms to continue fostering this style of collaboration. Everybody agreed that the surprisingly short response times stimulated the interaction, led to better insights, and yielded faster actions by management. And yes, management is joining these creative sessions. The option to drill down to the original PoS data at item level, at any time, helps users develop assumptions about observed facts. The relatively short development time of this project was very encouraging.

The same applications have to become available to larger numbers of users in order to support everyday work. This means not only fast response times are a necessity, but also the capacity to execute multiple queries simultaneously.

5.2

Financials

As stated before, all aggregations for reporting, whether statutory or managerial, are always calculated from the line items up. Should we want to change the structure (hierarchy) of, e.g., the P&L statement, we just have to rerun the statement with a different hierarchy definition. We can change hierarchies, calculations,

create subaccounts, etc on the fly, or even have multiple versions in parallel. Consider changes in the organizational structure, or acquisitions during the fiscal year or simulations of these – all of this is now possible on top of the recorded financial transactions without changing transaction programs or, even worse, changing materialized aggregates with inevitable downtimes. Without changing the hierarchies, certain parts of the organizational structure can be carved out by applying a simple filter. The enterprise system does not hinder the management anymore to take action and reorganize the company when needed.

⇒ **structural changes to the performance reports are possible at any time**

Most of the financial reporting which was migrated from the transaction system to the Business Intelligence (BI) system, solely for performance reasons or the availability of better reporting tools, can now be brought back, since the performance is beyond any expectation and all the reporting tools (SAP and non-SAP) are available on the HANA platform. The Online Analytical Processing (OLAP) functionality from the SAP BW is now integrated in the HANA OLAP engine. The time scaling in the reports can vary from days to months, from quarters to years – any time period of your choice. Comparisons over time, currency conversions, and hierarchy processing are reusable functions in the Business Function Library (BFL) of HANA. The extensive library of SQL views exploiting the virtual data models makes it relatively easy to compose a new report or to develop an analytical application. For the presentation Fiori,

Business Object tools, the SAP partner solution from the company SAS or any other toolsets can be used. The data works best when inside of HANA so important advantages such as central access control remain possible.

⇒ reporting with variable time scales

The time to run all the reports for a month, quarter, or year-end closing is close to minutes, and the closing of the books is solely determined by the time the accountants and auditors need to verify, judge and adjust the books. The closing process is modeled as a series of automated and manual tasks assigned to specific task owners. Some steps in the closing process are dependent on previous activities or require manual input. A financial closing cockpit helps scheduling the process and provides an overview of the completion. Just imagine what it means that you can make adjustments and see the result in the consolidated numbers in an instant. Since in the closing process, some steps will be repeated over and over again, the intelligent business object-aware caching of result sets will come in handy – the response time for a rerun, now including the latest changes made to the financial data, has been brought to sub-seconds. The trick is that only the data in the delta store is processed, while the result for the main store is cached until delta store and main store are merged again. Think about the software industry, where a large portion of the transactions happen in the final days or hours of the period.

⇒ **significantly faster period closing**

Even if data has to go into a data warehouse for cleansing and transformation reasons, the speed with which we can transfer data from other systems into a HANA-powered BW is several dozen times faster, and the latency between the transactional S/4HANA and the BW is close to zero.

Account Displays

Auditors love the fact that all aggregations are now virtual, which means they are not materialized in the database. If you are authorized, any aggregation algorithm can be added on the fly, as well as any search can be specified ad-hoc and the results retrieved in seconds. Especially interesting is the ability for authorized users to search through textual comments that accountants can attach to the line items.

Another example for more flexibility is the account display of open and closed items. Via search we can identify, for example, all accounts belonging to a certain customer and pass the list of these account numbers as parameters to the SQL statement for scanning all accounting items for these accounts together. After sorting the result set by account number and joining it with the account master data, we have a report on all accounting transactions with the various subsidiaries of a large business partner. This report can spread across multiple legal entities of the reporting company, or be run for a single one. A slight variation shows all accounts for a given city, industry, or any other qualifying account attribute (see Figure 5.5). The elegance of these reports (applicable to many other business objects) is striking. Navigation through the data model does not follow hierarchical structures any longer, but scans all accounting line items first. This change in the flow of the algorithms cannot be mentioned often enough.

Data Reduction

All redundant tables have been removed, all indices except the primary key dropped, and all materialized aggregates replaced by calculations. The remaining footprint for the actual data is only 2–2.5% of the original if we apply actual/historical partitioning to the financial data. This dramatic consolidation of the data footprint is striking. But more importantly, any control effort to check the completeness and correctness of the multiple versions of the accounting line items becomes unnecessary, for there is only the central truth.

⇒ **footprint of the actual data is reduced to 2–2.5%**

⇒ **unprecedented flexibility in applications**

The zero response time approach works, but sometimes we have to use conventional wisdom to improve a system. Here is a small example of how a feature from the current accounting systems finds its application in the new design. It has always been convenient to keep all accounting entries in one table, yet we kept separate index tables in addition. Now, we separate the Accounts Receivable (AR),

Accounts Payable (AP), the Material Ledger (ML), General Ledger (G/L), and others in separate tables, and build a union of result sets when appropriate. This decreases the table size for AR and AP, which reduces the scan efforts and leaves database capacity for other activities. The effect is measurable, since these two accounting areas constitute a large part of accounting overall.

Dashboards for Finance

The speed of HANA allows for sophisticated dashboards where key figures are calculated instantly, and whenever possible, presented in a graphical manner. These key figures can give the users an instant overview of the business areas of concern. The date indicates when the key figure was calculated last. The key figures are interactive, and a drill down through multiple levels of detail down right to the transaction data helps to analyze the facts.

Graphical Analysis of P&L Results

The full use of graphical tools like SAP Lumira or other visualization assets makes the data exploration more efficient (see Figure 5.6). It is important to note that all this happens directly on the transaction data in real time. There are no more delays, of any kind, between data recording and sophisticated data analysis.

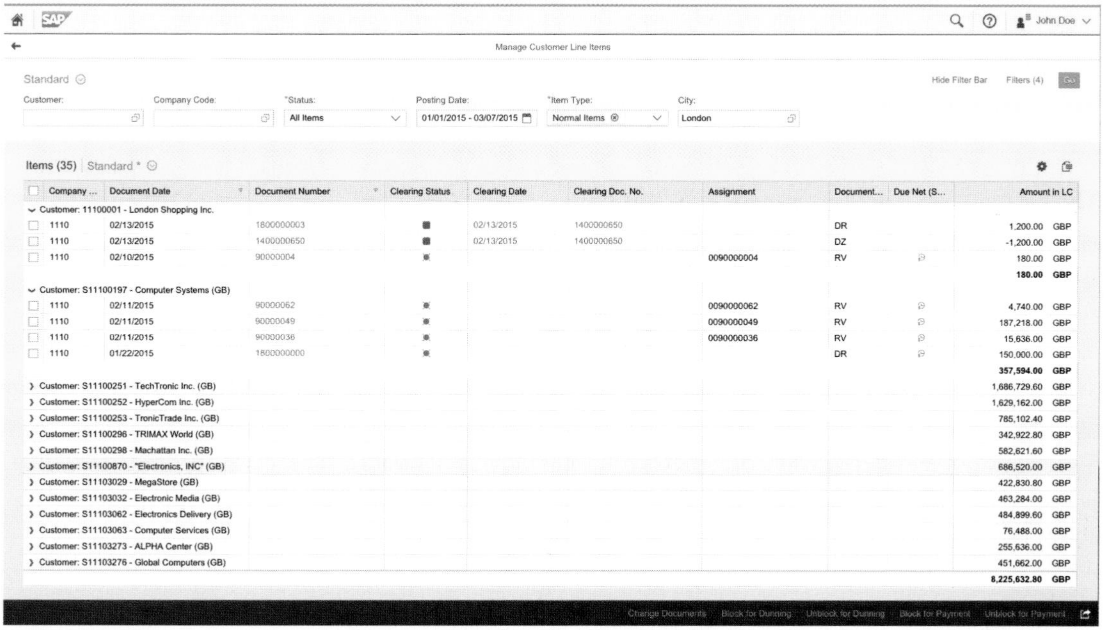

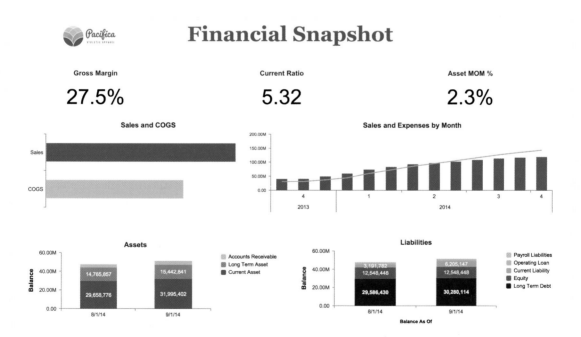

FIGURE 5.5
The Manage Customer Line Items app

FIGURE 5.6
Financial data rendered with SAP Lumira

The creation of the P&L statement does not take much more time than to display an account, and it is now a real-time transaction. The application remains fully interactive, which means the user can drill down to the highest level of granularity and choose the number of periods for comparison, such as a year in days or the last eight quarters. There is complete flexibility because everything is calculated from the line items upward. Several P&L positions can be selected for a graphical comparison. An experienced user is able to detect irregularities, surging costs, sudden reversals, and many other interesting facts. The response times are always sub-second, and a real intelligent dialogue develops, which will inevitably lead to a constant push for even better interaction design and presentation, or even new functionality. This is only one example of future finance applications which do not just produce static results anymore, but stay interactive and assist the user in data analysis. The interactive character of the application – a departure from the old purpose of documentation to present insights as a starting point for actions – makes the whole system so much more valuable. Figure 5.7 shows a screenshot of such a P&L analysis. The next level of development will show automated analysis, where the application presents already interesting data constellations based on machine learning.

Cash Forecast

One of the early showcases of the potential of HANA has been the cash forecast (see Figure 5.8).

What took hours on a traditional database to run, now runs within a few seconds. It is the convenience that matters. Together with other financial analysis applications, such as overdue reports, days outstanding, and the receivables and payables forecast, we always know the actual financial situation of the company. Volatile currencies or other risk factors can be included in the forecasting.

Simulation and Forecasting

Every company generally knows which external factors can influence their financial results in the future. Some refer to these factors as key performance drivers, and monitor them carefully: currencies, energy costs, labor costs, transportation costs, regional effects, raw material costs, etc. Again working in the P&L statement on the line item level allows us to predict future business based on the current actual and planned data for the year, modified by the selected performance drivers. Different levels of complexity are feasible, and response time is no longer an issue. Companies are building very sophisticated simulation applications as bespoke systems using business data, but want to protect their domain knowledge. The HANA Predictive Analysis Library (PAL) with its statistical and predictive algorithms such as time series analysis, regression, and classification builds the basis for these applications (see Section 8.1: Enterprise Simulation).

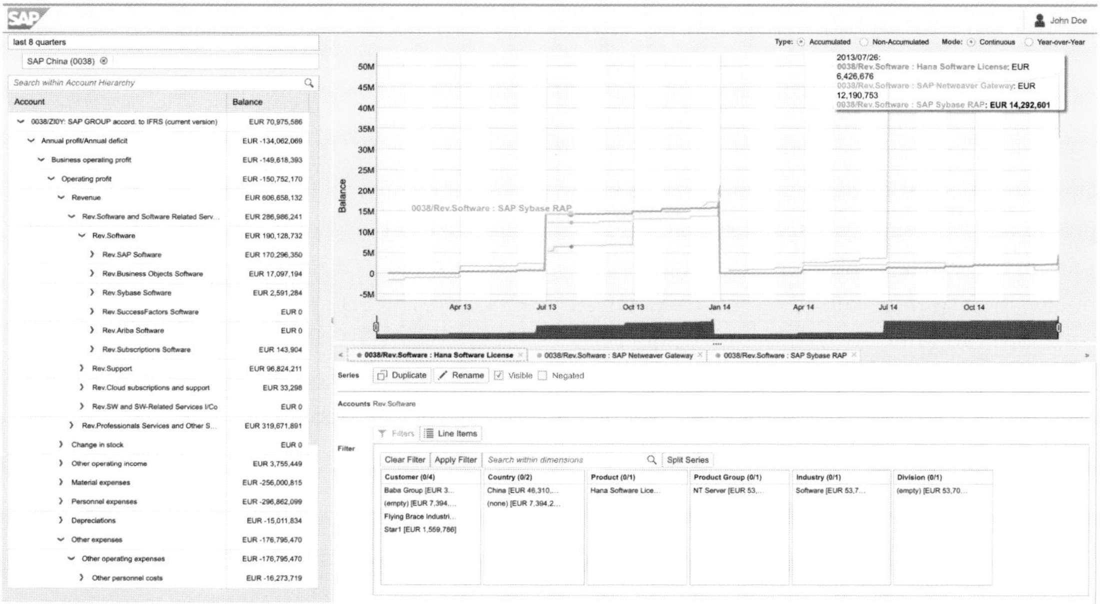

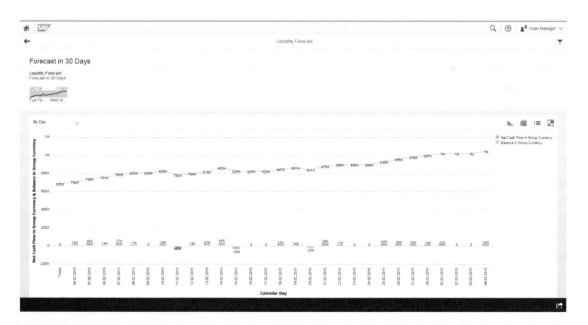

FIGURE 5.7
Graphical analysis of P&L statements

FIGURE 5.8
The cash forecast: the first showcase of the potential of HANA

To bring OLAP technology to transactional processes is changing the way work is done. Enterprise systems become an adviser and information source, instead of only a system to keep records. After the great hopes of the early seventies and the undeniable successes in the nineties, a new wave of smart applications will become part of the core business systems.

Cost Center Controlling

The primary costs are now a simple view of the G/L line items. Any check whether all accounting items with a cost center coding made it to the cost center controlling application is superfluous, the data is identical per definition. If there is any doubt, the cost account/cost center report will always equal the cost center/cost account for any period, and this proof will take only a few seconds.

Based on the planned data and the actuals, the system calculates an estimate of the cost for the period end. Potential cost overruns are identifiable early enough to take action. The simplified data model and the ability to work on line item level enables us to include commitments like open vendor orders or new hires, and analyze their impact on the period before they appear in the accounting system. Dashboards and graphical displays assist the cost center manager and the controllers.

5.3 Logistics

Similar to financial systems, logistics systems possessed many redundant data structures that had been introduced to improve response times and can be dropped in HANA. The simplification potential found in logistics is even larger than in the financial applications. In this section, we discuss inventory management, Available-to-Promise (ATP), Material Requirements Planning (MRP), material valuation, and logistics analytics. But first, let us understand how much more important simplicity is in logistics than in financials. The logistics system accompanies the physical processes of a company. These processes are changing with ever growing speed, as the Internet brought fundamental changes to this industry. Business networks allow companies to stay permanently connected with their business partners, and adapt to changes nearly instantly. As a consequence, product configurations have become more and more flexible, and we have customized products replacing mass-standardized ones, with lean manufacturing and smaller lot sizes being the trend. RFID tags help to track the location of physical goods and enable optimization of the flow of these goods. We can use trend analysis to forecast the next few days or weeks to optimize logistics.

The Internet of Things (IoT) brings a whole new challenge with regard to data volumes. New legal requirements such as Governance,

Risk, and Compliance (GRC) standards require innovative and flexible reporting. This all leads to a drastic increase in logistics transactions in the Enterprise Resource Planning (ERP) system. Any slowdown in these systems has a direct impact on the physical reality and the performance of a company. Since the world is developing at this high speed, it is not only mandatory to increase the overall performance of a logistics system by a significant factor, but to simplify it by an order of magnitude in order to prepare for the incorporation of future business processes. Of all the developments of logistics in SAP, from R/2, R/3 to ECC 6.0, the most recent development is, by far, the most radical and the most valuable one. While the innovations in S/4HANA's financials are an important step into the future, the ones in logistics are a must in order to be able to compete.

⇒ **performance of the logistics system directly impacts the performance of a company**

Inventory Management

From the start of SAP, stock quantities were kept as aggregated data in various tables. Every stock movement touched one or two of these preaggregated totals in the database system. To get a detailed report of the stock movements for a certain material, the application had to access up to 24 aggregate tables. HANA speeds up the calculation of central inventory numbers; for instance, we measured a speedup by a factor of 900 in the calculation of the opening stock, the stock changes, and the closing stock for a selected period of the past. Whereas previous systems had to access two database tables including the Material Document (MSEG) table, more than seven aggregate tables, two master tables, and at least five configuration tables, the on-the-fly aggregation of HANA accesses merely five columns of the MSEG table. This is only one example, but a good indicator of what the simplification and speed of HANA can do. As in financials, the new logistics transactions no longer create locking situations by updating aggregate tables – there are no aggregates anymore. With this, there is a significant impact on the scalability of inventory management, and a gain in system throughput by a factor of 3.5 to 7 is measured in backflush accounting for the automotive industry. With the help of compatibility views, older programs can still access data as before.

Customers do not need to preaggregate stock movement data anymore in order to cope with performance restrictions. They can now post more frequently, and the new level of detail supports applications such as parallel processing of PoS data, goods receipts handling, and invoice verification.

Available-to-Promise

A major change takes place in the ATP component. ATP is a business function that allows users to check the availability of products, control delivery quantities, and propose delivery schedules. This function is no longer based on aggregates, but uses the actual stock movements and production planning line items. With an optimistic approach (it is assumed that the product is available), the blocking of products is reduced, and in the case of failing availability, the recent requests will be analyzed to resolve the shortcoming. To work on the highest level of granularity enables us to make smart recommendations for how to solve conflicts from a sales perspective.

Production Planning

Running Material Requirements Planning (MRP) now takes significantly less time, depending on factors such as complexity, number of levels, and number of variations. Even in more complicated scenarios, where variation handling is required, we observed an improvement by a factor greater than ten. Multiple runs per day are now possible, and the flexibility in production scheduling is improved. In some installations, this alone justifies the transition of the existing ERP system to HANA.

Another area of great success is the optimization of transportation planning algorithms. Going from several hours to a few seconds for a transportation planning run can change an entire logistics strategy, and improve product availability for the consumer. Here, HANA shows the advantages of its columnar data organization, completely in memory, for computation-intensive algorithms.

A new MRP cockpit supports the job of production planners. Planners can configure an individual launchpad for themselves; half a dashboard for monitoring the material flow, and half a menu of analytical and transaction applications. The cockpit automatically detects disruptions in the material flow, and points to the most critical issues. This evaluation of the material flow happens frequently, and is independent of the MRP run. One analytical application shows, for example, the development of the situation for the critical materials in the next two weeks. Each planner configures further the details of the information. In the next level of detail, all demands and supplies for a critical material are brought together, and the planner can evaluate the impact on customer orders. The system provides an assessment of the situation and offers different options to resolve it, including a simulation of the impact. These options are rated, and the planner can choose one to be put into action. Here, we see again how speed translates into more intelligence and an improved User Experience (UX). While these options have been offered before, there was no decision support for the planner and possible solutions were not evaluated with regards to cost and impact on customers.

Material Valuation

Material valuation determines or records the stock value of a material, which for discrete products is the stock quantity times the material price. Since inventory management is based on the transaction data and does not use any aggregates, the same approach was taken for the valuation. Multiple valuation methods for different accounting standards and currencies exist in parallel, such as based on a periodic price in the Material Ledger (ML).

5.4

Sales and Marketing

Most parts of the Customer Relationship Management (CRM) functionality will return to the core sales system in the end. The CRM data could be distributed, for example, company data to the core, and a sales representative's data to the Sales Force Automation (SFA) tools. SFA tools change rapidly and have characteristics similar to office systems or social networks – they need to draw information from all kinds of sources, and help build presentable information in the most comfortable way. A package of coupled applications, offered in a public cloud, is most likely the right answer. We can now combine various on-demand apps easily in a cloud environment because the interfacing technology is so much more robust, and the variations for different system software are reduced in the Cloud.

Impact of Big Data

Pipeline management, pipeline reporting, and predictive sales analysis based on direct and indirect unstructured information have all the aspects of Big Data. The power of HANA to combine unstructured data retrieved from different sources with structured data (customer, order history, etc) is unmatched, even if data is first captured in Hadoop or similar systems. Speed will become the dominant factor in CRM. Being able to handle Big Data in close to real-time will give companies a competitive advantage.

⇒ **Big Data analytics in real time**

In addition, marketing solutions typically have to work on a large amount of data. For example, a dramatic impact can be seen in call center and service applications. The combination of text and structured data processed together at a high speed improves the service quality for the customer.

Point of Sale Systems

While the traditional PoS system (the cash register) is still in place in physical stores, and will be for a foreseeable future, more and more transactions will happen in online stores. These vary from simple shops, built from public cloud offerings, to sophisticated websites/stores, built completely according to the owners' specifications. Running the more sophisticated ones in the HANA Cloud makes sense if data integration (even running through services) with central order processing systems is a vital part of the application. The higher the volume, the more HANA has to offer. In any case, PoS systems running in the HANA Cloud will integrate with any meaningful on-demand application, from credit card handling to information about consumer behavior. Typically, these systems consist of a collection of applications sharing an application platform. Using the HANA platform for the next generation of PoS systems provides major advantages. Many online shops lack speed, and the consumer can often choose an alternative vendor if kept waiting. The small data footprint supports deployment in the Cloud, and allows the system to keep all data in memory.

In the case of a multi-channel sales organization, it still makes sense to bring all PoS data with the highest level of granularity into one system for data analysis. For retail companies, these systems can be huge. HANA has done a tremendous job in reducing the data footprint, and has provided data scientists and marketers with unprecedented response times, while working through hundreds of billions of line items.

Real-Time Sales Data Analysis

Analyzing sales data is one thing. Even with nearly zero response time for detailed investigations of sales by product, country, region, customer type, etc, we are still looking at the past. With HANA, it is possible to monitor in real time all sales activities regardless through which channel (direct sales, physical shops, Internet shops, or indirect sales) the data arrives. We can analyze the correlation between sales promotion activities and the market reaction. Since HANA

can process text data, for example, from publicly available social network feeds, and index this data, structured and unstructured data can be analyzed in one system using OLAP, predictive analytics, and statistics in R. This combination of data from different sources allows for new, previously unavailable insights.

Speed is not a solution in itself, but it enables us to explore the data and to ask a second or third question in order to gain a better understanding of what is going to happen in the marketplace. As long as we stay in the flow of human thought processes and respect our limited attention span of a maximum eight seconds, we will stay engaged and our curiosity will drive the investigation.

Comparisons between similar products in different locations, variations in pricing, impact of local promotions, etc can be examined during the window of opportunity, meaning that variables like price, promotion, or advertising can be changed, and we can observe the impact of our measures in real time. This opens up a completely new world of managing sales activities on a global basis, faster than any of the competitors still stuck in conventional data processing.

The first customers using HANA in sales clearly stated the unbelievable value of speed, predictive analysis, and modern presentation of information using a wealth of visualization tools. The ability to view changes in price or the effects of promotional materials during the day shifts the focus from a period and campaign-based business to a true real-time business.

Particularly interesting is the analysis of PoS data in real time. We will not only be able to correlate data with external events, being reported by the news, discussed on social media networks, or simply related to weather, but an early reaction is likely once we have understood the impact. Aggregated data cannot do the job, we need the day, even the hour when the sales happened for the detailed analysis. The option to permanently monitor the sales activities in real time, comparing sales regions, products, and probably returned goods gives us the opportunity to completely reorganize the customer relation regardless what sales channel was used. In Section 10.2: All Key Performance Indicators at a Glance, we will discuss the Point-of-Sale Explorer app, which allows a broad spectrum of users to have an overview of thousands of products.

The periodic reporting is still there, only it comes instantly at any variant of period end. Redundant briefing books via spreadsheet and presentation software should be history (see Section 5.1: The Boardroom of the Future).

Customer Segmentation for HSE24

Home Shopping Europe 24 (HSE24) is one of the leading names in modern, multichannel home shopping. The innovative mail-order company keeps people up to date on the latest trends with its channel brands HSE24, HSE24 Extra, HSE24 Trend, and its online shop. Its interactive TV, online, and mobile platforms are systematically networked, and it also offers extensive smartphone, smartTV, and tablet applications. The HSE24 lifestyle brand provides its customers with a shopping experience around the clock. Addressing over 41 million households in Germany, Austria, and Switzerland, HSE24 had an active customer base of 1.5 million and delivered 11.2 million packages in 2013.

Each year, the HSE24 product range features more than 20,000 products (most of them exclusive) in the areas of fashion, jewelry, beauty, and home & living. The company generated net sales of €551 million in 2013 and has more than 700 employees, with another 2,200 people working for HSE24 at call centers and logistics companies. Since 1995, HSE24 has been continuing on a path of sustainable growth [Mul14].

Currently, there is an exponential increase in customer data. More and more channels arise and need to be integrated with existing customer information. Online platforms, mobile apps, and social media generate Big Data which impedes home shopping networks such as HSE24 in their approach of selling live. Faced with these challenges, HSE24 looked for a new, integrated, and comprehensive view of their customer data and an efficient multi-step campaign management. A solution should allow them to address customers via their preferred channel in a personalized fashion and react quickly according to changing buying behavior as well as company strategy.

We present the HANA-based Audience Discovery and Targeting (ADT) tool as a solution allowing HSE24 to better understand their customers. Based on HANA, this tool stores customer data at the highest level of granularity and supports Big Data analytics in real time.

Identify Target Groups

Targeting the customer segments that would be most likely to respond to specific campaigns is a key task for efficient campaign management. In the past, the Business Intelligence (BI) group was required to analyze this information, a process that could take a week or more.

Instead, marketers should be able to create target groups for campaigns without technical knowledge. For this reason, a new segmentation solution must offer intuitive visualizations of results. Moreover, it must be able to process all HSE24 customers and their sales data, which is stored in various SAP and non-SAP systems, in real time. Particularly important for marketing experts is the ability to investigate customer data in an explorative way in order to be able to gain new insights into customer preferences.

SAP ADT enables marketers to play with their data.

Audience Discovery and Targeting

SAP ADT powered by HANA offers a solution for HSE24. As part of the SAP Customer Engagement Intelligence (CEI) marketing solution, it enables marketers to perform the required analysis independently and flexibly create target groups for campaigns. Marketers can query the data to detect customer buying patterns based on demographics such as age and location. The analysis results are then represented in graphics and charts, allowing for discovering patterns in the data. Marketing staff can drill down into millions of customer records in seconds using recent data from a variety of sources, e.g., from HSE24's SAP Customer Relationship Management (CRM) system or their non-SAP data warehouse.

Figure 5.9 shows an example of SAP ADT. On the left side, this screen offers a list of customer attributes and key figures, which in this case have been extended and adapted to the needs of HSE24. On the right side, ADT offers the segmentation model, and below, a preview of the customer distribution for the selected segment. In this example, a marketer analyzes customers who have bought cooking articles according to various criteria, e.g., gender and age, which are selected from the menu on the left hand side. Each segment shows the number of customers who fit the criteria. Set operations such as intersection can be applied to segments to allow flexible data analysis. The preview at the bottom not only offers a graphical representation of the data, but also allows the marketer to select the customer base for the next segment to be created.

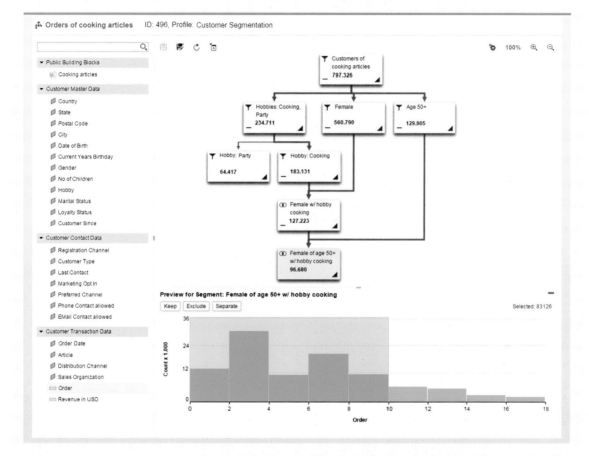

FIGURE 5.9
SAP Audience Discovery and Targeting (ADT) allows users to segment customers and create target groups. This example selects female customers, age 50+, with cooking as a hobby (segmentation graph on the right side). The chart at the bottom presents the distribution of the target group with respect to the number of ordered cooking articles.

The chart shows the distribution of customers according to the number of orders for cooking articles. Different preview types are available, e.g., a map for the selection of countries or a tag cloud for hobbies. Once the desired target audience has been identified, a target group for a new campaign can be created directly.

This application enables users to be business intelligent. SAP ADT enables marketers to play with their data, nurturing their creativity in order to find new insights.

Granular and Parallel, for Any Kind of Data

The basis of a strong segmentation solution such as SAP ADT is the availability of all data on the finest granularity, offering the highest flexibility to end users. HANA enables real-time, on-the-fly computations such as joining data from different tables, filtering according to given criteria, counting the number of orders for each customer, grouping customers according to their number of orders, as well as date calculations.

Customer segmentation relies on providing flexible selection capabilities combined with comprehensive previews. Especially the latter benefits from HANA's high scan speed and its ability to apply massive parallelization, e.g., when counting the number of orders for customers.

Finally, HANA is able to work on any dataset and offer comprehensive data replication tools.

HSE24 loads relevant data in near real-time from SAP CRM, and only once a day from their data warehouse. In the future, this setup can be further simplified by running SAP CRM and SAP ADT on a single HANA instance, making periodic data loading unnecessary.

Sentiment Analysis

In addition to working with HSE24 on customer segmentation, the next step for strengthening the segmentation solution within companies is to incorporate more accurate information on individual customers. A highly informational source for understanding customer interests and trends lays in public social network posts, emails sent to company accounts, and written complaints. This data can be merged into the marketing data universe with SAP Social Contact Intelligence.

This data enrichment enables marketers to understand current trending topics and most active social users. With this information, they can directly schedule follow-up actions such as creating target groups or having collaborative discussions.

The sentiment analysis loads any unstructured text data into HANA. The built-in text analysis engine interprets and tags the free text. These tags are then structured into different types containing mentioned topics, company and product names, and also their related sentiments. Even the intent of the post can be understood, e.g., if it is an urgent request for support.

Figure 5.10 shows the sentiment analysis for the search term "SAP." With this tool, a marketer is able to identify increases in the absolute number of tracked posts from the past eight days, just by looking at the upper middle chart in which green areas highlight positive sentiment trends. In the tag cloud below, the identified topics with the highest increases in traffic are displayed, while their letter size indicates the absolute number of posts, and the color highlights the rate of increase. This shows marketers where they need to become more active to either strengthen a good run or smooth down differences. On the left side, filters allow marketers to further narrow down their analysis. Moreover, the marketer can immediately take action by creating post groups in the upper right corner. This leads to the creation of target groups for subsequent campaigns or collaborative activities. All the analyzed data further strengthens the customer profiles of ADT.

understand current trending topics and most active social users

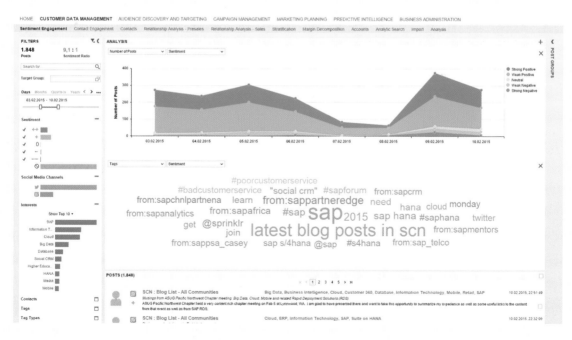

FIGURE 5.10
Analyzing social opinion about the search term "SAP"

5.5

Analytics and Planning

All analytical applications benefit from the changes to the data model in HANA. Operational reporting is now reintegrated into the transaction system and runs on the live data in real time. The impact on system operations (data transfer, synchronization, redundancy, latency, footprint, etc) is substantial, and additional functionality can be developed much faster. This is particularly important for the development of industry-specific analytical solutions and their implementation on top of standard systems.

Business Intelligence and Enterprise Performance Management

How can we commercialize the digital assets of a company? Analysis of Big Data is a prerequisite for discovering the business value of digital assets to a company and its potential customers. SAP Analytics utilizes HANA as the real-time data platform with the ability to scale for large datasets, SAP Advanced Analytics with its predictive algorithms, and SAP Lumira for true self-service data exploration and visualization.

According to Gartner, "SAP has continued investing in new capabilities with a number of potential differentiators. For instance: SAP Lumira's advanced self-service data preparation and infographics features; smart data discovery, by integrating Predictive Analytics (SAP's advanced analytics platform acquired from KXEN) and Lumira; governed data discovery, by integrating SAP Lumira with SAP BusinessObjects Enterprise, and integration with SAP HANA." [SHS+15]

SAP Analytics also extends to the world of modern financial processes. It is a paramount objective of SAP Analytics to deliver the unified experience of a combined world of Business Intelligence (BI) and Enterprise Performance Management (EPM), i.e., read-only processes for reporting and analytics, and the realms of planning, budgeting, and forecasting. Business users with a focus on BI encounter capabilities for reporting, dashboarding, and analytics. As soon as planning functions are needed, users will find them only a click away. This is facilitated by the coherent reuse of key algorithms and data structures which occur both in reporting and planning. Consequently, planning, analysis, and simulation can be freely combined. We will go into more detail in Section 7.2: Financial Planning in the Cloud, in which we describe the topic of financial planning in the Cloud, and in particular, the SAP Cloud for Planning offering.

All analytical features can be used in any planning context. For example, forecast data can be generated from a prediction algorithm in HANA. HANA can generate a number of scenarios to meet a certain Key Performance Indicator (KPI) target or to inform a user which data points correlate strongly with the accounts involved in the planning. Advanced logic, such as intercompany elimination, is defined only once and is immediately available for both analytics and planning.

Conversely, all planning features become usable in all analytical contexts. Plan data is of major interest when analyzing data, as everyone needs to know how the business is doing in comparison to its goals. The system allows the user to compare different plan versions or even company-wide scenarios without the need for additional modeling. Any database view in HANA, such as the view of all revenue transactions, can be used for planning and simulation – without moving or changing the data. Hence, true what-if analyses become possible in any report.

A good example of algorithmic reuse is the set of functions that handle the effect of currency changes on the business results. While generating new values for a plan version (e.g., for revenues, costs, or margins) on a corporate level, a certain status of the currency values in the foreign subsidiaries is assumed. If the plan version is a function of comparable actual values of the past, currency conversion to the corporate currency has to be applied. Whenever we later compare the plan against new actuals, we can determine the effect of currency changes in the meantime, i.e., the share of the plan/actual difference that is due to currency development.

With the convergence of BI and EPM, we create software that is much closer to the work of the people in various business departments. SAP Analytics aims for a closed loop experience in business information management. We have discussed BI, SAP Advanced Analytics, and EPM, however, to close the loop additional components are necessary. We integrate tools and technologies to move data in real time from various heterogeneous sources into one single dataset. By doing so, we need to ensure data quality and master data coherence. All of this is bundled in the Enterprise Information Management (EIM) package which is part of SAP Analytics. Many corporation-wide KPIs can only be defined on such a combined and harmonized dataset. In addition, we provide tailored applications for given analytical scenarios that grow from a critical mass of requirements in a given industry or Line of Business (LoB) for which it is worthwhile to create a packaged application to address use cases.

All of this is built on HANA and its incarnation in the Cloud, the HANA Cloud Platform (HCP). The HCP is an important asset for all of SAP Analytics as it provides engines for Online Analytical Processing (OLAP), planning, text analysis, geospatial analysis, graph data, and predictive algorithms out of the box. SAP Analytics can concentrate on delivering business value, and we can offer an integrated experience such as in Figure 5.11. Consequently, the SAP Analytics experience is attractive for business users and IT professionals at the same time.

Demand and Supply Planning

SAP Advanced Planning and Optimization (APO) on HANA benefits from the consolidation of all technology components relevant for supply chain planning into HANA. The APO application code was optimized to utilize HANA's performance benefits by reducing the number of database accesses for read operations. APO on HANA introduces HANA-based fast search and fact sheets for selected master data objects. This enables fast access to detailed master data information with contextual navigation.

For finite supply chain planning and simulation of transactions on networks of orders, APO uses an object-oriented storage, which is an integrated part of the HANA platform. APO demand planning and the extraction of APO data for central analytics uses the embedded Business Warehouse (BW). For usage with HANA, the BW-based data model was simplified and accelerated by replacing multi-dimensional BW cubes with data store objects. In addition, the redundant storage of precalculated totals for extraction was eliminated.

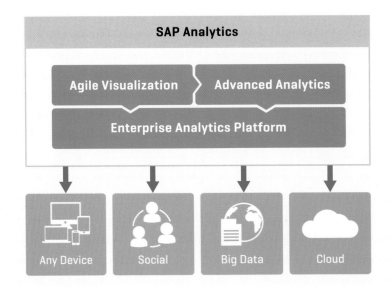

FIGURE 5.11
The integrated experience of SAP Analytics

performance improvements from 40% to 60% for demand and supply planning

Productive customers in mission-critical implementations show performance improvements from 40% to 60% for end-to-end demand and supply planning processes as well as an improvement greater than 50% for backorder processing runs. Planning and exception management data can be displayed three to five times faster with optimized read-operations. APO on HANA customers also have optimized self-built reports and can now retrieve important end-user information ten times faster.

SAP Supply Chain Info Center (SCIC) is an extension to APO providing HANA-based supply chain analytics. SCIC enables decisions based on the latest information, and trade-offs are considered in the full context of the KPIs and business performance objectives. Visualization through HANA Virtual Data Models (VDMs) (see Section 2.2: Advanced Features) allows instant analytics on large amounts of data with user-defined selections and filters. Customers have proven that they can eliminate several steps in their planning process and, along with the performance improvements mentioned above, transform a weekend planning job into a daily job. This provides not only more recent and accurate planning results, but also leads to a much faster response to demand and supply changes.

Sales and Operations Planning

In a Sales & Operations Planning (S&OP) process, different roles within a company work together on a mid to long-term plan for the demand, supply, and financial sides of the company. Typical areas which must be addressed are projections of sales numbers, margin optimizations, decisions concerning product mix, and identifications of bottlenecks on the supply side.

To make S&OP precise, the planning process takes into account company-wide information, with detailed data for individual entities such as customers, products, sales channels, manufacturing locations, and resources, as well as sometimes bills of materials for critical components. As a result, the S&OP process becomes a Big Data problem in which often a billion or more individual numbers need to be stored and many more computed on the fly to derive key figures based on complex formulas.

The SAP Integrated Business Planning (IBP) solution contains a S&OP application which utilizes HANA's high calculation speed to change the way users perform the S&OP process. S&OP is now an interactive, consensus-based planning process. The S&OP application can be used directly in planning and review meetings. It enables users to interactively simulate and create planning scenarios or make changes to the entire company's plan on the fly. For example, if a user changes the current aggregate level, the results are available in seconds even though the calculation performs several hundred million steps. The system is also able to keep track of KPIs to compare different versions of the plan.

HANA enables plan changes to be propagated down to the most detailed planning level (along supply chain network relationships) and aggregates the results back for the planner in a single interactive step. For these calculations, S&OP types of key figure relationships often require a lot of join operations because S&OP spans multiple dimensions. Dependencies between key figures can be twenty levels deep. All S&OP calculations of this type are performed by the HANA calculation engine.

Key figure calculations can be modeled specifically for each company – S&OP is not usually standardized across companies. At configuration time, the system generates performance-optimized code for the HANA calculation engine. At runtime, the calculation engine optimizes further, only computing what is required to answer the specific user query. No aggregates or intermediate results are persisted.

Warehousing

Warehouse supervisors have now up-to-the-minute operational visibility across all warehouse processes and KPIs, including the possibility to instantly drill-down to the single order line. HANA enables analytical

Supervisors are empowered to quickly identify issues, take actions, and employ corrective measures.

monitoring of one million order lines in less than a second in order to continuously update the KPI dashboards and provide timely visibility of changes in the processing status of any single business document. We can now provide real-time insight into operational performance across outbound, inbound, and internal warehouse processes. Moreover, we can compare actual performance of shifts with the performance of previous shifts. This empowers supervisors to quickly identify issues, take action, and employ corrective measures.

Extended Warehouse Management (EWM) benefits considerably from the Big Data processing and predictive analytics capabilities of HANA. Large distribution centers face volatile order volume flow and need to plan and handle between 500,000 and 1,000,000 order lines on a single day. Labor Demand Planning (LDP) uses HANA to calculate and visualize upcoming workloads as the basis for efficient warehouse resource planning and assignment.

With HANA, it takes less than 800 milliseconds to calculate the required daily demand for warehouse resources based on 500,000 order lines of data. It is now also possible to predict unknown, short-term future demand. Here, LDP uses HANA's Predictive Analysis Library (PAL) algorithms to forecast the expected workload based on historical data and detected trends, trend distribution, and peak days. The algorithm can read over 300 million records and provide a forecast for the next ten working days in 1.1 to 2.1 seconds.

Transportation Management

An example of the benefit of analytics for logistics is in transportation. Transportation managers now have real-time visibility of freight operations, as well as an overview of KPIs and event monitoring. Reporting is performed across transportation requests and orders, costs, revenues, resource utilization and discrepancies, as well as delivery statuses (see Figure 5.12).

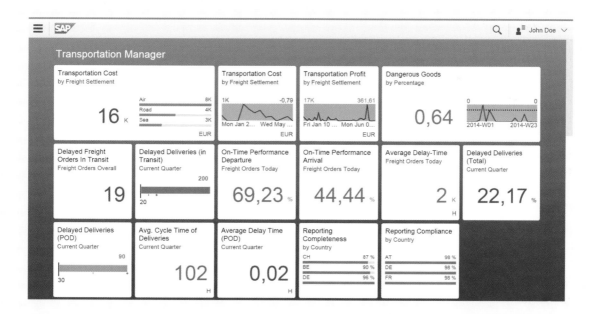

FIGURE 5.12
Real-time reporting for a transportation manager

FIGURE 5.13
Transportation Resource Planning: empty container planning

In the area of transportation costs, the complexity of the core data model was significantly simplified. The number of database tables involved for processing charge line data was reduced from ten to two. This results in up to eleven times fewer read or update accesses during runtime. As an example, a mid-size scenario with 50 charge lines and four dimensions per charge line results in a reduction from 2,700/1,350 (read/write) accesses, distributed over 10/5 (read/write) database tables, down to 250 accesses in total, distributed over two database tables. This simplification leads to huge performance improvements in all areas of transportation management, as charge calculation is a central tool supporting freight and forwarding order management processes, i.e., for collective invoicing of carrier services.

Besides improving existing functionality, new functionality has also been developed taking HANA design principles into account to avoid predefined aggregates and redundant data structures. One area possessing new functionality is contract management. Here, the consumption of freight contracts is calculated on the fly, leveraging transportation management transaction data. Another example solution, Transportation Resource Planning, directly accesses the transaction data of transportation management to determine and visualize the number of empty containers available at specific depot locations (see Figure 5.13). There are no aggregates and there is no need to update aggregates with operational transportation stock movement data. Planners now have complete visibility across container usage, as well as availability for the fast identification of demand/supply imbalances. Resulting stock transfer orders for empty containers are seamlessly incorporated into the operational processes of transportation management.

We have rethought data processing for the area of transportation planning and optimization. Specifically, transportation network determination takes a significant amount of time for larger scenarios. Transportation lane determination runtime increases dramatically as the number of locations/zones increases and the transportation network setup becomes more complex. In two test scenarios (based on real customer data for North America), HANA enabled us to reduce the runtime to determine connections between locations (transportation network/lane) from over twelve minutes to less than two seconds. This makes it possible to reduce overall planning runtime, improve planning optimization quality (saved preprocessing runtime can be used for optimizer runtime), and optimize more complex scenarios with a greater number of locations in one planning run. The economical and ecological benefits reduce transportation costs and carbon emissions through optimized freight demand consolidation across business units and regions, as well as inbound and outbound movements.

5.6 Lessons Learned

For many people the stability of the new Business Suite was a complete surprise. How can such a revolutionary change work so well? The answer is short – simplification. Every single aspect of the change contributed to a more simplified system. If there are no further updates of data structures, nothing can ever go wrong with updates. Everybody who has ever built a piece of software knows that the updates are the most critical part of the application. If there are no redundant data structures, we do not have to make sure that these structures are in sync with each other. In the event that an application algorithm fails, we just correct it and rerun it – nothing more is required. Imagine the invoicing index, a trigger to produce invoices at the right date. It had to be created, processed and then deleted. Now, it is only a scan for the customer orders due to be invoiced, which is only a fraction of the previous system complexity. It is the speed of HANA which allows us the opportunity to shave off complexity wherever possible, and the resulting simplicity guarantees stability.

PART THREE

KEY BENEFITS OF HANA FOR ENTERPRISE APPLICATIONS

In all areas of business, we welcome speed and simplification while lowering costs and raising customer satisfaction. While integration of business processes had excited the business world in the nineties, a focus on incremental improvement and cost reduction slowed subsequent progress. It is with the modern ability to provide services via the Cloud that new life has been brought to the industry of enterprise applications. The true value of these applications comes from their focus on the simplicity of consumption in an era of growing complexity for traditional on-premise solutions.

In this part, we present the key business benefits of HANA through a selection of real-word examples. The presented examples are either projects done with a single customer tackling a focused problem (e.g., accelerate a certain process, reduce maintenance costs, etc), or a proof of concept conducted by SAP or one of its partners with no particular customers involved (e.g., a new navigation concept). We will always begin with an introduction to the general setting of a particular business domain preceding the following sections:

 POINT OF VIEW

This section states either the job to be done or the pain point of a target end user from the considered business domain.

SOLUTION

This section demonstrates how SAP, often together with its co-innovation partners, solves the problem for the end user.

ENABLEMENT BY HANA

This final section describes how the features of HANA enable the provided solution.

In the first chapter of this part (Chapter 6: Leveraging Design Thinking in Co-Innovation Projects), we introduce Design Thinking and SAP's Co-Innovation Initiative approach with which many of the examples were realized. Each of the remaining chapters introduces a key business benefit of HANA, demonstrates it with a selection of examples, summarizes the findings, and discusses the outcomes.

Outline of this Part

opportunities and make good decisions, we need to have access to the right business information at the right time. This allows us to act quicker, and thus, reduce costs. This is true for planning, customer and supplier interaction, and reporting and analysis. For all of these areas, the possibility to explore and analyze real-time data can completely change how we run our business.

CHAPTER 10 **QUALITY TIME AT WORK** HANA allows us to simplify not only how we run applications, but how we use them as well. The improved response times make an instant change. Using the new UI technology, SAP Fiori allows us to redesign most transactions and reports. This technology is platform-independent and provides a uniform User Experience (UX) across mobile phones, tablets, and desktop computers. Using SAP Fiori, the development process becomes much leaner and faster, enabling new enterprise solutions to be developed quickly and tailored to the needs of distinct roles within a company. This empowers users by providing them with all information needed to get their job done.

CHAPTER 11 **SOLVING THE UNSOLVABLE** All around us, Big Data challenges are encountered in nearly all areas of business. Whether evaluating terabytes of sensor data from the Internet of Things (IoT) or analyzing billions of messages from social networks, such challenges are a common occurrence. HANA delivers the speed necessary to process such huge amounts of data. It possesses the built-in functionality needed to analyze and understand this data through its predictive algorithms and sentiment analysis. Building on these results, HANA performs simulations and what-if analyses, and by doing so, guides businesses into the future.

CHAPTER SIX

LEVERAGING DESIGN THINKING IN CO-INNOVATION PROJECTS

Throughout history, disruptive innovation has influenced our behavior and resulted in progress. In order to drive disruptive innovation, SAP conducts co-innovation projects which help understand what their customers desire and address these needs. To facilitate the success of a co-innovation project, Design Thinking is applied to develop an idea and prototype a solution exploiting the technical advances of the HANA platform. Below, we will describe SAP's Co-Innovation Initiative and how Design Thinking enables co-innovation, and then conclude with two examples of the co-innovation model.

6.1
SAP's Co-Innovation Initiative

It is necessary for SAP to be closely engaged with its customers in order to provide customer value. To further this pursuit, SAP has recently added a new set of services enabling an innovative collaboration model with partners: SAP's Co-Innovation Initiative. Within traditional development projects, topics are typically chosen by SAP and customers are approached to collect their feedback. In contrast, co-innovation projects are initiated by a customer challenging SAP to solve a particular business problem they are currently facing. These co-innovation projects are SAP's approach to investigating how its technological innovations can be used to solve the customers' real-life problems.

To facilitate collaboration, these projects follow a well-defined co-innovation model. First, the customer and SAP come together and discuss the pain points as well as possible solution enablers. This allows the SAP team to develop a high-level understanding of the value that such a solution would provide to the customers. Following this, the initial prototypes are built and a customer-specific business case is created to ensure that the customer's objectives are properly addressed.

These initial prototypes are used to decide whether and how to continue the project. If the project continues, it will then go into the co-development phase. During this phase, SAP and the customer jointly develop a detailed prototype of a possible solution. At the end of this phase, SAP contributes by building a customer-independent business case that helps to identify the appropriate next steps for the project. Based on the customer's feedback and the developed business case, SAP and the involved partners decide how to proceed. Either SAP will develop a standard product based on the prototype which will then be available to the customer, or SAP will stop its work on the project and grants the customer usage rights. In case the customer wants to continue working with SAP to develop a customized solution, the project will be handed over to the custom-development teams.

Co-innovation not only gives customers technical solutions for their problems, but also provides the complete experience of identifying the problem, ideating, iterating, designing, and executing. All these phases happen hand in hand with potential end users validating each step from concept to consumption.

In the context of such co-innovation projects, intensively applying Design Thinking has proven to be very helpful. Therefore, SAP considers it to be a vital factor for the success of such joint projects.

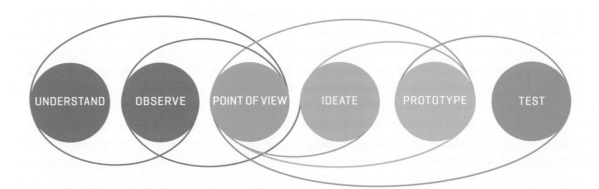

FIGURE 6.1
A Design Thinking process model [PMW09]

6.2

Design Thinking as a Co-Innovation Enabler

Design Thinking is an iterative and interdisciplinary team-based approach. Through the use of this approach, collaboration and creativity are fostered with an emphasis on user-centric design and simplicity. Design Thinking helps to identify problems, find solutions, and drive creative innovation with the help of design-related working materials such as sticky notes, whiteboards, and toy bricks. The process begins with ethnographic methods, such as interviews and observations. Afterwards, sketching and prototyping techniques are used to receive early feedback and iterate quickly. In contrast to most classical software engineering methods, user studies are conducted throughout the development process – not just in the later phases.

Several descriptions of the Design Thinking process exist, with different names for its phases and methods; there is no single rigid process, and several techniques from different variations are typically applied depending on the tackled problem. At Stanford's d.school, the HPI School of Design Thinking, and SAP, the following phases are commonly used (see Figure 6.1):

Understand
Analyze the (complex) design challenge, discuss its various understandings, and focus on one or a few of its core aspects.

Observe
Use qualitative user research, such as interviews and observations, with end users. The goal is to understand why users behave in a certain way and what their objectives and needs are.

Point of view
Create a common point of view for research results within teams. In this phase, the different observed perspectives have to be consolidated and each team needs to come to one point of view.

Ideate
Produce a large pool of ideas based on the point of view as the basis for prototyping and testing. Wild ideas – potential seeds of innovation – are encouraged. In the end, teams should decide on the most promising ideas they want to prototype.

Prototype
Build a prototype to get early feedback on whether an idea has potential or not. Depending on the availability of time, granularity, and purpose, prototypes can be made out of paper or toy bricks, interactive click-through programs, or role-plays.

Test

Collect feedback on the prototypes from future users to validate their potential. Testing is also a good way to discover which aspects of the prototype are problematic and how to circumvent these.

The phases of the Design Thinking process are not performed sequentially, but rather iteratively. For instance, depending on the results of the Test phase it might be necessary to return to the Prototype phase to improve the tested prototype (if the idea seems to have high potential), to the Ideate phase (if new ideas based on the feedback are needed), or even to the Observe phase (if new insights are necessary).

Besides the iterative process, an important aspect of Design Thinking is the diversity of team members. This diversity facilitates co-innovation by enabling the inclusion of different perspectives. During their academic and professional lives, people develop expertise in specific domains and adapt a certain way of thinking that is relevant for their work. However, other perspectives help people see problems from different angles and may reveal interesting and unanticipated aspects. Therefore, having diverse team members from different areas is an integral requirement to Design Thinking. This helps ensure that Design Thinking remains user-centric and provides real value for the end users by creating new solutions, rather than simply adding functions and features to existing solutions.

In the following section, we show how SAP uses their co-innovation model and Design Thinking techniques to develop a new solution for real estate managers.

6.3

Co-Innovation in Action: The Real Estate Cockpit

SAP has been challenged with a business problem that many facility management units face nowadays: "How do we enable facility and real estate managers to easily analyze the efficiency of their buildings?" Following the co-innovation model, SAP organized a Design Thinking workshop to investigate this question within diverse teams including participants from Siemens Building Technologies, the SAP Innovation Center, SAP User Experience (UX) Design Services, and SAP Global Facility Management. Within these teams, the participants identified and prioritized a variety of different functional features and roles. Based on these findings, the teams built paper prototypes in a first iteration. The prototypes captured the most important Key Performance Indicators (KPIs) and functional requirements of the end users as shown in Figure 6.2a.

In a second iteration, after analyzing the results of the workshop and performing additional user research, the SAP UX Design Services team came up with a generic management cockpit which can be adjusted to the needs of different types of users. The team designed preliminary versions of this cockpit (see Figure 6.2b). They then verified these

versions with the initial workshop participants and additional potential end users from SAP Global Facility Management and Siemens Building Technologies.

In a third iteration, the SAP UX Design Services team refined the cockpit concept and the visual design (see Figure 6.2c). Meanwhile, a team of software engineers from the SAP Value Prototyping department started developing a prototype (see Figure 6.2d).

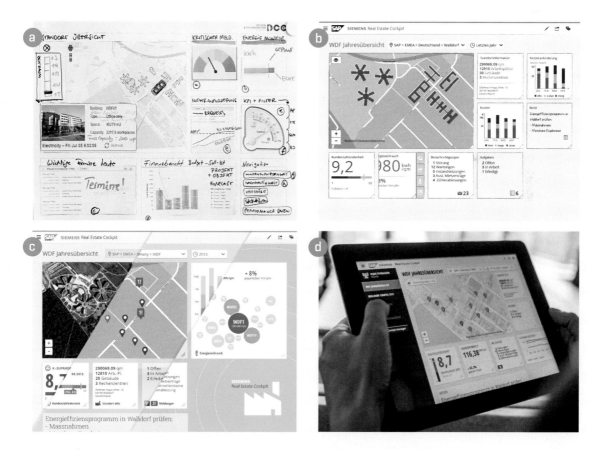

FIGURE 6.2
Paper prototype, wireframe, visual design, and final prototype of the Real Estate Cockpit

The resulting prototypical solution provided predefined cockpit dashboards for key user groups such as facility managers, asset managers, portfolio managers, and sustainability managers. The dashboards enabled the users to quickly obtain building data, as well as business facts, to evaluate the buildings' overall efficiency.

The positive experience of the participants was acknowledged several times. "Before we started the project, we talked to real estate managers who were skeptical due to the massive amount of information and the underlying complexity. But if you extract meaningful data from underlying systems, you get a picture of what you actually need to manage buildings. Real estate managers now have all the information they need in a much faster way, and many KPIs can be combined for easy access to all the relevant data," says Peter Marburger, Head of Sales Energy Efficiency – Siemens Building Technologies.

"The [Real Estate Cockpit] co-innovation project is a successful example of how co-innovation supports stakeholder alignment and incorporates user needs," reflects Marion Fröhlich, Senior Strategic Design Consultant at SAP.

The solution was so well-received by the involved partners that it is now being implemented by Siemens and SAP Global Facility Management. Furthermore, due to the overwhelming positive feedback, the solution is now also being turned into an SAP standard solution at the SAP Innovation Center.

6.4 Co-Innovation in Action: Medical Research Insights

Healthcare is one of the aspects of our lives that will be impacted the most by information technology in upcoming years. Trends such as personalized medicine and new medical technologies will change the way diseases are diagnosed and treated. In this operating model, medical decisions, such as therapy approaches to be applied and drugs to be used, are tailored specifically to the individual conditions of the patient. This also poses new challenges for physicians and healthcare professionals, all of whom must consider many individual pieces of information from different sources and collaborate in interdisciplinary teams to make the right treatment decisions.

Comprehensive cancer centers can be seen as the vanguard of this development. They combine proven treatment approaches, cutting-edge research, and preventive medicine in order to provide cancer patients with the optimal support for overcoming their disease. However, just like running an innovative business, the volume and variety of data that needs to be processed for providing comprehensive cancer care is immense. Yet, while business users can choose from a wide variety of powerful Business Intelligence (BI) software products, physicians, particularly oncologists, require analytics software that is tailored to their specific needs.

The goal of the Medical Research Insights (MRI) project is to provide such a specialized solution. In cooperation with the National Center for Tumor Diseases (NCT) in Heidelberg, the project team designed, developed, and delivered a new analytics software for clinical research. This solution allows physicians and researchers to access patient data from various systems in real time with a single interface to improve cancer research.

Personas

Researcher

"To answer hypotheses, create studies, and publish research results, we need a system that identifies patients with similar cancer characteristics."

Needs
> identify topics for clinical studies
> select relevant patients by finding accessible tissue samples
> analyze medical correlations with data mining on large, anonymized patient cohorts to verify research hypotheses

Challenges
> acquired data often suffers from low quality which hinders the selection of patients and testing of hypotheses
> the sharing of study results requires a reliable de-identification of patient data to ensure that individuals are not traceable

Physician

"In order to optimally treat our patients, we have to include them in clinical studies investigating new ways of treatment and better medication."

Needs
> access patient data and histories
> reference scientific studies and clinical trials that are relevant for a specific patient

Challenges
> medical data is often diverse, hard to compare, and redundant
> therapies and treatments require massive documentation efforts
> sharing knowledge with other physicians is difficult

FIGURE 6.3
Personas of a researcher and a physician

Background

The National Center for Tumor Diseases (NCT) Heidelberg was founded as an alliance between the German Cancer Research Center (DKFZ), Heidelberg University Medical School (HUMS), the Medical Faculty of Heidelberg, and German Cancer Aid (Deutsche Krebshilfe). NCT has rapidly grown into a comprehensive cancer center of excellence and is uniquely positioned to benefit from the wealth and depth of DKFZ cancer research and the Heidelberg biomedical campus.

The NCT's mission is to foster interdisciplinary oncology for an optimized development of current clinical therapies, and to rapidly transfer scientific knowledge into clinical applications through a comprehensive concept of translational and preventive oncology. NCT offers state-of-the-art interdisciplinary strategies for diagnosing, treating, and preventing cancer and for conducting innovative clinical trials. Currently, approximately 8,200 newly diagnosed cancer patients are seen at NCT and more than 12,200 patients were treated at the NCT outpatient therapy unit per year.

NCT has implemented an innovative Precision Oncology Program (NCT POP) as a center-wide master strategy. NCT POP coordinates all translational activities and focuses resources towards individualized cancer medicine, including patient-oriented strategies in genomics, proteomics, immunology, radiooncology, prevention, and early clinical development. The overall aim is to provide individualization of treatment decisions according to the specific needs of each patient by reaching across traditional disciplines and academic departments. NCT employs more than 300 scientists, physicians, and biomedical staff. The two Personas in Figure 6.3 summarize their work, needs, and challenges.

Clinical trials are vital for improving cancer treatments and making them available to patients. The NCT's programmatic effort in clinical research is focused on dedicated clinical cancer programs with particular clinical expertise and research impetus in Heidelberg, demonstrating significant trial activities and overseeing large patient cohorts. NCT currently enrolls 14% of all treated patients into more than 300 open clinical protocols, approximately 70 of which are investigator-initiated trials driven by local researchers.

Laborious Handling of Medical Data

YESTERDAY, the creation of a new clinical trial required researchers to manually identify a list of inclusion and exclusion criteria that match specific cancer types and patient characteristics. With this list, they asked a trusted IT employee to find suitable study participants by examining several medical databases. In doing so, the IT employee had to translate the colloquial formulation of the criteria into multiple SQL queries, then summarize the various database responses into spreadsheets, and finally return the results to the researchers. This entire process was usually quite laborious and required several weeks to complete because researchers and IT employees needed multiple iterations to clarify requirements again and again.

In the past two decades, there has been a move towards evidence-based medicine, which demands the systematic review of clinical data as well as treatment decisions based on the best available information.

Large datasets need to be processed in order to derive robust statistical evidence. This data includes not only structured content such as clinical trials and documentation of diagnosis codes, but also unstructured, freely-formatted text documents such as doctor letters and medical publications. The following two examples show two specific workflows that can particularly profit from an automatic analysis of all this medical data with a fast and scalable solution:

Patient cohort analysis

In order to evaluate the effectiveness of a new chemotherapy drug, a cancer center invites some of its patients to participate in a clinical trial. However, these patients need to match certain criteria regarding defined characteristics such as the type of diagnosis, the previous treatment history, and their age. After completing the clinical trial, physicians need to evaluate the tested drug by comparing the trial cohort's response to the response of a similar cohort that received traditional drugs. Yesterday, these tasks – finding patients that match the required criteria and comparing the participants' responses – involved going through patient records manually, collecting the relevant information, and consolidating it in spreadsheets or other mediums.

Personalized medicine

The goal of personalized medicine is to tailor medical decisions, treatments, and even drugs specifically to the individual patient – and not only to the diagnosis. This can significantly improve the patient outcome, which increases the healthcare provider's reputation. However, analyzing a huge and diverse amount of (partly unstructured) medical data in a short time – as it is required to be able to quickly make solid, personalized decisions – is not possible with yesterday's technologies.

Supporting these and other medical processes with a dedicated IT solution requires three major challenges to be met which relate to the huge amount of data: ensuring usability, being able to conduct real-time analyses, and having a flexible data model.

First, tools for healthcare providers must be tailored to the needs of medical staff. These tools should support them to perform their tasks as efficiently as possible. The process of querying a large amount of data, both structured and unstructured, and the representation of the results must be easily understandable for domain experts such as physicians and researchers. For software vendors, usability is of growing importance as the influence of chief medical officers on decisions regarding IT purchases is increasing. However, developing expert tools that are simple to use is a major challenge.

Second, until now, the distinction between relevant and irrelevant data required an incredible amount of time spent on paperwork and data compilation. However, making decisions as soon as possible to be able to help the patients as fast as possible requires analyses to be conducted in significantly less time.

Third, medical information is composed of diverse data that is spread over different systems. For example, a patient's medical history is distributed across visited physicians' offices and hospitals, while published cancer studies are scattered across several libraries. For this reason, it is often very laborious for physicians and researchers to find the information they are looking for. In order to solve this challenge, a new and flexible data model is needed to cope with the diversity and large volume of medical data.

Accessing Medical Data from Various Systems in Real Time

TODAY, researchers can create new clinical trials on their own with a few clicks. With the help of Medical Research Insights, they can immediately analyze clinical data to accelerate cancer research, find suitable patient cohorts, and match patients with the best clinical trials. Only in the case of a specific data source not being available, an IT employee has to assist with the integration into Medical Research Insights. However, this has to be done only once and the effort required is limited due to a flexible data model and the features of HANA.

Medical Research Insights (MRI) was developed in partnership with the NCT. It gives instant access to clinical information and patient data from multiple sources such as clinical information systems, tumor registries, and biobank systems. Based on a generic and flexible healthcare data model, physicians and researchers can analyze and

The MRI project has received a lot of international recognition and has won several awards.

most up-to-date information available. With HANA, SAP is currently in a unique position to develop and deliver truly game-changing IT systems for healthcare providers. SAP technology enables MRI to handle large amounts of diverse data, to analyze it in real time, and to visualize results in less than one second.

The following figures (Figure 6.4 to Figure 6.7) illustrate the features of MRI and describe how they help to overcome the challenges of yesterday. They also show the intuitive User Interface (UI) and how it helps medical staff work efficiently.

visualize complex datasets in real time, as well as filter and group patients according to a wide range of characteristics.

By selecting a subset of patients using filter criteria, a patient cohort can be created. This can be edited collaboratively, exported for further analyses, or compared with other cohorts regarding different metrics such as number of patients, age, and treatment response. MRI also supports researchers in the evaluation of clinical trials by presenting the treatment success in a concise and understandable way. Furthermore, the tool offers a holistic overview of each individual patient's medical history in a graphical timeline that can support medical staff during patient visits or tumor boards and makes it easy to access information on all levels of detail at any time. Designed to serve different roles, MRI replaces time-consuming and manual processes, streamlines workflows, and highly increases efficiency across various medical teams.

In the clinical environment, it is crucial to base decisions on the most accurate and

A Distinguished Project

The MRI project has received a lot of international recognition and has won several awards for its approach, technology, and design:

> recognition by the White House Office of Science and Technology Policy, jointly with other SAP healthcare activities
> partnership with the American Society of Clinical Oncology (ASCO)
> Strategy and Research Professional Notable Award at the Core77 Design Awards
> German Design Award 2015 in the category "Interactive User Experience"
> two SAP internal Awards – the HANA Innovation Award in the category "Social Hero Innovator," and the One Team Award 2014

Patient Analytics – Filter and Analyze Patient Data for Patient Cohort Building

MRI enables medical staff to securely analyze clinical data to identify research options and facilitate matching patients with the best clinical trials, e.g., for the analysis of multiple cohort studies.

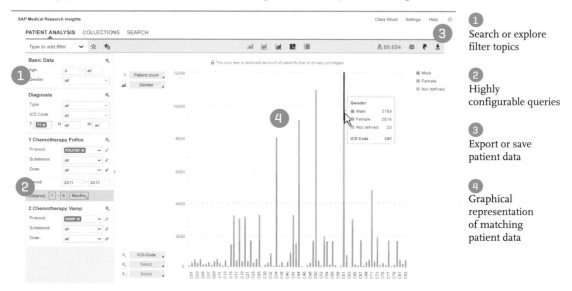

1 Search or explore filter topics

2 Highly configurable queries

3 Export or save patient data

4 Graphical representation of matching patient data

FIGURE 6.4
Patient Analytics

Search through different data sources
MRI gives instant access to clinical information from multiple sources for comprehensive cancer care.

Highly configurable filtering
Medical staff can filter patient data according to different attributes. Sequential and temporal conditions, as well as positive and negative filters, can be included.

Bookmark filter settings
Filter settings can be bookmarked for later use.

Flexible graphical representation
The visualized charts can be rearranged to match the user's needs, e.g., switching the axes or using different chart types.

Patient Collections – Collect, Edit, and Manage Patient Data for Patient Cohort Analysis

Patient collections can be used for patient cohort analysis, patient-trial matching to identify patients as potential participants for trials, or the validation of research hypotheses for new trials.

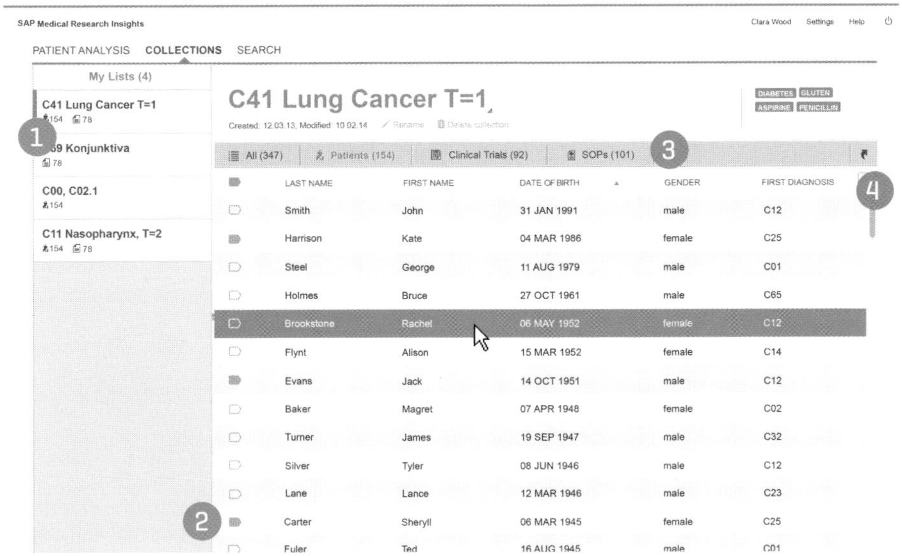

FIGURE 6.5
Patient Collections

1. Patient and mixed data collections
2. Mark reviewed patient data
3. Filter by document type
4. Customizable columns for different tasks

Enrich patient list with related information

Collections consisting of patients who, for example, have participated in a clinical study, can be used to store related scientific articles or references to clinical studies. This keeps all information in one place and enables a quick overview of related information.

Collection management and extension

Collections can be saved and extended directly from both the patient analysis view and from the search results view. A typical workflow could look like this: a researcher identifies a patient cohort by filtering different clinical criteria, saves the collection, looks for related content using the search functionality, and then enriches the set of patients with the new content.

Sorting and drilling into items

The items of a collection can be sorted according to arbitrary types. The details of each item can be accessed with a single click. This is highly efficient because MRI stores patient data on the highest level of granularity, without any materialized aggregates which would hinder flexibility.

Patient Timeline – Immediate Access and Comprehensive Data of Patient History

Medical Research Insights offers an overview of each individual patient's medical history, supporting medical staff during patient visits and tumor boards.

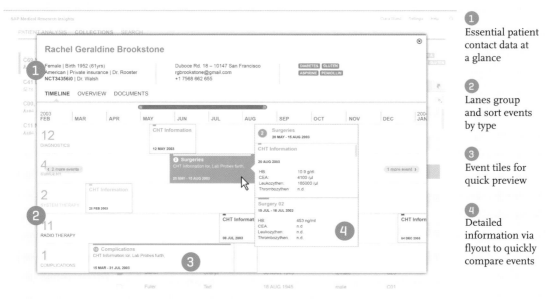

1 Essential patient contact data at a glance

2 Lanes group and sort events by type

3 Event tiles for quick preview

4 Detailed information via flyout to quickly compare events

FIGURE 6.6
Patient Timeline

Graphical timeline

The graphical timeline view makes it easy to access information on any level of detail. It helps physicians to obtain an overview of the patient's medical events, understand the relations between different events, and investigate further details.

Flexible healthcare data model

Thanks to a flexible healthcare data model that can handle a wide range of data types, patient data from many different sources can be accessed simultaneously.

Patient Evaluation – Visualizing and Exploring Complex Patient Datasets in Real Time

Physicians and researchers can immediately explore the selected patients' information through different interactions, settings, and chart types, as well as analyze complex datasets from a wide range of medical events – all in real time.

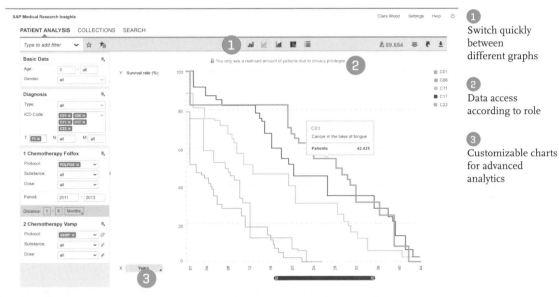

FIGURE 6.7
Patient Evaluation

1 Switch quickly between different graphs

2 Data access according to role

3 Customizable charts for advanced analytics

Visual previews of clinical data

The system offers various visualizations to preview and explore available data before filtering down to narrower sets of results.

On-the-fly survival statistics

Physicians and researchers can directly compare the survival times for different patient groups (e.g., those receiving different treatments) using the common Kaplan Meier estimator [KM58].

The Impact of Medical Research Insights

Real-time analytics software saves time by straightforwardly obtaining meaningful data

Medical Research Insights (MRI) is a tool specifically tailored to the needs of physicians and researchers. The in-memory capability of HANA allows for the processing of massive amounts of data in real time. The volume and variety of information which needs to be processed for providing comprehensive cancer research is immense. The ability to access and analyze data in real time not only saves time, but also drives improved personalized care for the patients.

Intuitive and appealing interface designed for different roles with easily accessible key functions

The uncluttered interface features a workflow-oriented structure which is intuitive to use, requiring minimal training. The User Experience (UX) is based on an interaction model that optimizes usage without losing functionality. In addition, the interface suits not only the experts, but roles from all knowledge levels. For example, less frequent users can explore clinical data via menu-based access, while experts can use intelligent context-based search bars to search for specific data.

New scalable product solution which is customizable for different cancer centers and hospitals

In response to an important demand of the healthcare sector, MRI was co-designed with medical experts from the NCT, where it is now in use and developed further. This SAP product can be easily configured for different cancer centers and hospitals by adapting individual elements, such as data sources and filters. This means that other institutes will also be able to obtain better research results and care for patients more effectively.

Streamlined workflows increase the efficiency among medical teams

MRI integrates different sources, streamlines workflows, and replaces manual processes to increase efficiency among various medical teams. At the same time, it limits access to information according to user status for maximum data privacy.

High-Performance Clinical Data Analytics

MRI is a native HANA application. The SAP Fiori-based presentation and interaction logic runs in a web application that communicates via HTTPS. The HANA platform connects with these applications and runs control flow logic on top of the database. All data-intensive processing and computation steps run close to the data, directly within the HANA database. An external application server is not needed.

This main architecture is already able to overcome two out of the three major challenges that hinder new medical use cases – usability and real-time analytics. SAP Fiori, in combination with Design Thinking, provides an ideal foundation for creating a UI which can be intuitively used by all medical staff, and the underlying HANA database supports real-time analyses.

The third challenge – dealing with the diversity of medical data – is more difficult. Doctor letters, medical publications, or biomarker test results are hard to analyze automatically. Their free-text, unstructured, and non-standardized nature mask the included valuable information. To extract the important patient parameters and store them for later analysis, MRI builds on HANA's text processing and a flexible data model that leverages the properties of a fast columnar layout.

*data processing and
computation in HANA*

Text processing

HANA provides a range of features for text processing that help to extract important information and identify relevant keywords, names, and facts of entities from documents. HANA offers a rich set of language modules, system dictionaries of predefined entity types, and linguistic models to derive new entities. Furthermore, the entire text processing can be customized, which is especially important for the medical domain with its very complex terminology. For example, automatic text processing needs to be able to recognize the fact and the difference between "evidence for a mutation in the ALK (Anaplastic Lymphoma Kinase) gene," and "we could not identify a gene mutation in ALK."

MRI builds on HANA's text processing engine and supports customizations required for the medical domain. Customized dictionaries list new entity types that, in turn, contain a standard form name and any number of synonyms. Additional extraction rules, written in the Custom Grouper User Language, allow for formulating patterns that match tokens by using a literal string, a regular expression, a word stem, or a part of speech. With these customizations, HANA automatically performs linguistic processing, i.e., language and encoding identification, segmentation, case normalization, stemming, tagging, and entity and fact extraction. The statement, "we could not identify a gene mutation in ALK," in a doctor letter can be analyzed and MRI is able to identify "ALK" and "gene mutation" as the relevant entities, extract the "not," and interpret this sentence as test result "ALK negative." Finally, the normalized representation of this result is stored in a flexible data model which can handle and analyze a wide spectrum of medical facts.

Entity Attribute Value (EAV) data model

To support the dynamic addition of patient parameters as extracted from free-text documents of varying medical data sources, MRI uses an EAV data model for its flexible foundation. Each fact of a patient is stored in a simple database row of the form [patient ID, fact type, fact value], e.g., [33, "primary tumor," "lung cancer"]. This physical representation of the data effectively enables storing new patient parameters without

changing the data schema. Through MRI's graphical interface and the corresponding dynamic SQL query builder, the user is shielded from this underlying model, and instead works transparently with medical concepts such as patient filters and attributes.

While this data model design is observed in other clinical study data management systems, its implementation on a traditional database has led to the low performance of analytical queries. This is because the execution of these queries requires a number of joins and self-joins to put

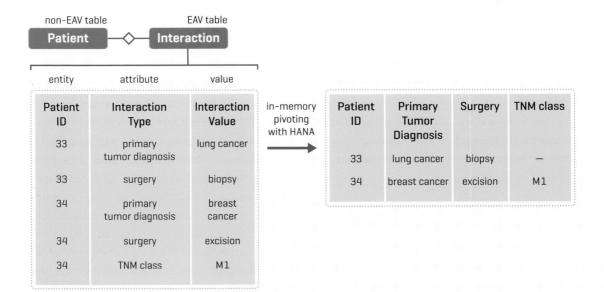

FIGURE 6.8
Simplified data model for healthcare data in the form of EAV tables. The attribute columns are pivoted in memory to later calculate aggregates dynamically. In the example, data on three medical attributes (diagnosis, surgery, and TNM (tumor class) classification) stored in an EAV table is pivoted into a format with one column per attribute.

the results back into a normalized form and to discard irrelevant patient entries according to the requested cohort. This costly operation, known as pivoting, reconstructs rows from the flexible EAV data model. When implemented in HANA, however, real-time analysis is made possible through two key mechanisms: first, by processing the analytical queries in main memory, and second, by employing a columnar layout for the EAV tables. In order to enable users to dynamically explore the clinical data, analytical queries are generated on the fly according to the configuration of the EAV tables. With the help of the query builder, the patient cohort is converted into a corresponding SQL statement.

These queries can require multiple joins over the tables for entities such as patients, interactions, and their details. In the back-end, the EAV tables are pivoted into a normalized form in main memory as presented in Figure 6.8, only rebuilding rows with the columns needed for the specified patient cohort, and subsequently, calculating aggregates. In this way, MRI enables exploratory analysis of clinical data with sub-second response times.

> Within the area of personalized medicine, there are several new opportunities for innovative software solutions.

Personalized Medicine for Each Single Cell

TOMORROW, Medical Research Insights will enable researchers to access data from even more patients and larger information sources, such as genomes and proteomes. Users will be able to upload their own data results, complement clinical trials of other researchers, and share their findings with other institutes. With this broad data source, it will be possible to create additional insights by analyzing only existing studies, without collecting new data or recruiting patients.

While MRI is already being used to accelerate cancer research, upon approval as a medical device in the near future, MRI will improve healthcare in general. Patient cohort analysis will be accelerated by automatically identifying ideal matches of patients with relevant clinical trials. Physicians will immediately find the best possible therapy based on specific patient characteristics directly following a cancer diagnosis. Through transparent patient histories and individualized therapy plans, workflows will be made more efficient. With such time-saving features, medical staff will have more time to focus on the important things – the treatment and care of patients.

In the more distant future, personalized medicine will be increasingly important. This does not only include the thorough analysis of all patient data as shown with MRI, but also the integration of more complex data sources such as the human genome. This shift in medical practices is driven by a deeper understanding of the biological causes of diseases such as cancer, for which research has revealed a number of molecular pathways that determine the behavior of tumor cells. The activation of these pathways depends on the individual genetic makeup of the patient, and thus, can explain why a specific chemotherapy protocol might be successful for one patient yet ineffective for another. Within the area of personalized medicine, there are several more opportunities for innovative software solutions. For example, the SAP Innovation Center has already developed two other prototypes with HANA which support the following use cases:

Targeted therapies

Targeted therapies take the cell biology of an individual patient into account. An example of this is the drug "Trastuzumab," which targets a specific genetic factor that causes uncontrolled cell proliferation in certain aggressive types of breast cancer. In cooperation with the Max-Planck-Institute for Molecular Genetics and Alacris Theranostics GmbH, the SAP Innovation Center has developed a prototype for a Virtual Patient Platform which can simulate drug effects on the cell biology of individual patients and with this, aid the development of new targeted therapies.

Molecular diagnostics

In order to apply targeted therapies successfully, diagnostics are needed that identify relevant biological factors such as genetic mutations contributing to cancer. Molecular diagnostics often rely on building large reference databases containing biological data, such as the genome of many patients, and running statistical analyses on this data. Such usage creates an ideal use case for the HANA platform. In the project Proteome-based Cancer Diagnostics, in cooperation with Freie Universität Berlin, the SAP Innovation Center implemented a HANA-based analysis pipeline for protein concentration data that could lead to a diagnostic test for lung cancer. This project also initiated the development of a HANA-based tool that allows researchers to model such statistical analysis pipelines interactively in order to make their development more efficient.

CHAPTER SEVEN

REDUCTION OF COMPLEXITY

H

ANA simplifies the way we build and use enterprise systems. The integration of Online Transaction Processing (OLTP) and Online Analytical Processing (OLAP) was the first step for this simplification; the removal of predefined aggregates was the second. Now that all redundancy has been removed from the system, enterprise applications can finally run analyses directly on the transaction data.

Outline of this Chapter

SECTION 7.1 **DATABASE FOOTPRINT REDUCTION IN ENTERPRISE SYSTEMS** When SAP switched its Enterprise Resource Planning (ERP) to HANA, the database memory footprint dropped from 7.1 terabytes to 1.8 terabytes. With S/4HANA, this footprint will further decrease to 0.8 terabytes in total, including an actual data partition of 0.2–0.3 terabytes. This translates directly to a sharp reduction in the Total Cost of Ownership (TCO) as all backup, replication, and recovery processes are reduced accordingly.

SECTION 7.2 **FINANCIAL PLANNING IN THE CLOUD** By taking the SAP Cloud for Planning as an example, we explain how HANA and the Cloud can reduce the complexity of standardized financial planning and analysis processes. This flexible tool can be applied for individual and ad-hoc financial planning sessions at any time and at any place. Furthermore, the cloud deployment helps companies ensure low costs, adaptability to changing business needs, and scalability for any size of organizational unit.

SECTION 7.3 **CELONIS PROCESS MINING – A NEW LEVEL OF PROCESS TRANSPARENCY** For a global company with over 300,000 employees and a revenue of around €75 billion, it is crucial to understand how its business processes are executed in order to remove inefficiencies and deviations from the standards. To give companies insights into their processes, the startup Celonis developed its solution Celonis Process Mining based on HANA, providing analyses for variant processes, root causes, and throughput times, as well as continuous process control.

SECTION 7.4 **SIGNAL DETECTION IN BIG MARKETPLACES** Every day, million users generate billions of data entries in online marketplaces. In a co-innovation project with one of the world's largest online marketplaces, SAP created a tool that can simultaneously monitor over a million performance metrics to detect anomalies affecting the health of the marketplace. Forecasts can be created and compared to the actual data in real time.

7.1

Database Footprint Reduction in Enterprise Systems

Over the past decades, enterprise systems could achieve reasonable performance only by maintaining materialized aggregates. Consider the example of financial transactions: enterprise systems did not only insert raw debit and credit transfers between accounts, but also updated aggregated balances. On-the-fly aggregations were simply not feasible because calculations of balances would require the expensive summation of all transactions. These high costs are caused by the large amount of transaction data that is spread over multiple disk pages, which in turn, requires many slow Input/Output (I/O) operations to fetch this information. For this reason, materialized aggregates had to update results during each transaction and so allowed fast access to predetermined analyses. However, the maintenance of these materialized aggregates adds unnecessary complexity and overhead to enterprise applications. Furthermore, the concept of materialized aggregates significantly increases the effort required for data insertion, consumes additional memory for redundant data, and limits flexibility with respect to answering arbitrary analytical queries. The former assumption that all materialized aggregates required for the majority of applications can be anticipated without slowing down the applications is outdated.

HANA overcomes these drawbacks by supporting the simplest approach: only recording raw transaction data. It allows for simple inserts of account movements without the need for updating materialized aggregates. Balances can be calculated with the summation of all account movements on the fly. This database is capable of running analytical queries of large enterprise systems directly on redundancy-free data. In consequence, HANA introduces a simplification which redefines the way enterprise systems are built.

With HANA, we only record raw transaction data without the need for updating materialized aggregates.

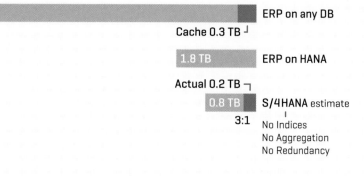

FIGURE 7.1
Database footprint reduction in the SAP ERP system by moving to S/4HANA

Lower TCO by Minimizing the Data Footprint

The Enterprise Resource Planning (ERP) system of SAP, a $22 billion company with over 66,500 employees [SAP14a], had a data footprint on disk of 7.1 terabytes with a database cache with 0.3 terabytes in main memory on a traditional database. SAP ran its ERP system on a single 80 core, 4 terabyte Dynamic Random-Access Memory (DRAM) server. When more reporting and analytical functionality was required, the company planned for a second server to replicate the most needed information from the ERP system. Both the original ERP system and a potential replica require certain IT needs to be met – amount of CPUs, disks, and main memory. As the costs for this setup depend on the size and complexity of the underlying data model, it is SAP's endeavor to reduce the data footprint, and thus, minimize necessary hardware.

💡 SOLUTION

Reducing the Volume of Enterprise Data

As a first step forward, SAP switched its ERP system from running on a traditional relational database to running on HANA. This switch already led to a significant reduction of the data footprint, as illustrated in Figure 7.1. The same

system as SAP was previously running, without any substantial changes in the application data, needed only 1.8 terabytes in HANA because of the columnar layout and its enabled data compression (see Section 2.1: Core Principles). With this, the system already experienced a reduction factor of approximately four. The next version of the ERP system, S/4HANA, features a system without aggregates, redundant tables, and excessive database indices. The elimination of all redundant data was estimated to reduce the storage demand below 0.8 terabytes. After a further split into actual and historical data partitions, the remaining storage requirement for the actual partition will be only 200–300 gigabytes. The historical partition is less frequently queried, and therefore, the data can be (but does not need to be) in main memory (see Section 2.2: Advanced Features). As a result, the actual partition is reduced to as little as 3% of the original ERP system. With the move to S/4HANA, HANA needs less main memory for its actual partition than traditional ERP databases used for caching. This is a remarkable reduction of the data footprint.

Notably, besides the reduction of the data footprint from 7.1 to 0.2–0.3 terabytes, all following backup, replication, and recovery processes are also accelerated. Moreover, the included systems for production, development, quality assurance, and backup require less storage and so can also be reduced. This also applies to cloud-based systems. Such systems benefit even more from a smaller data footprint.

⊶ ENABLEMENT BY HANA

Avoiding Redundancies and Compressing Data

There are four main reasons for the reduction of the data footprint of enterprise applications on HANA.

1. Columnar layout

In an in-memory database with columnar organized tables, only populated columns consume main memory. This especially reduces data footprint in standard software as no single customer is using all attributes.

2. Dictionary compression

Data is encoded with the help of dictionaries and stored as memory-efficient integers. Additional compression techniques for the numerical attribute vector which represents each column reduce the storage space even more. Especially in enterprise applications where many columns have few distinct values, these concepts can save much memory.

3. No aggregates and additional indices

HANA does not need redundant data. There are neither materialized aggregates nor indices that require additional memory beyond the primary key (and sometimes a secondary group key). Data is stored only on the highest level of granularity and all aggregations are computed on demand.

4. Data tiering

Enterprise data can be split into actual and historical partitions. While actual data includes current business information that is kept in memory, historical data is not required for daily business and so can be stored on disk.

All these features contribute to a reduced data footprint; yet, the real improvements come from the removal of all redundant data and unnecessary indices, with materialized aggregates replaced by database views and on-demand calculations. This simplifies the system tremendously.

7.2

Financial Planning in the Cloud

How successful is my business? Financial analysts working in the Controlling department must have an answer ready at all times. Understanding whether a business has been successful or not requires comparison of the results to the planned goals. In order to reach goals, planning and analytical activities have to go hand in hand in a process which is called the "plan – monitor – analyze cycle." This is increasingly important for smaller parts of organizations because they too want to establish individual planning and reporting cycles.

In this section, we describe SAP Cloud for Planning, a technology which reduces the complexity of standardized planning processes and data models in order to create individual, flexible, and intuitive planning and analytical processes. Based on HANA and running in the Cloud, SAP Cloud for Planning ensures fast adaptability to changing business needs and scalability to any size of organizational unit.

POINT OF VIEW

Ad-hoc Planning

Planning activities that only involve parts of an organization usually happen irregularly as business is steadily changing. For example, a research idea or a new potential project can trigger a planning process. A plan is not always built on regular time intervals – it is rather a management decision to do so. There is no standard data model available that fits this particular planning scenario. Often, this is a situation where businesses start using spreadsheets to build the plan. Even though this may work in the short term, the missing integration of plan data into actual company systems will lead to synchronization efforts in the long term.

⚲ SOLUTION

Simpler Planning with SAP Cloud for Planning

SAP Cloud for Planning is a solution that satisfies individual business financial planning and analysis needs. It comes as a ready-to-use planning solution that is purely focused on end user requirements and is based on a powerful cloud architecture that ensures important IT requirements such as availability, reduced maintenance effort, and lowered costs.

End-User Perspective

Figure 7.2 presents the web-based perspective of the SAP Cloud for Planning application. In the center, the main planning and analysis spreadsheet-like data modeling system can be

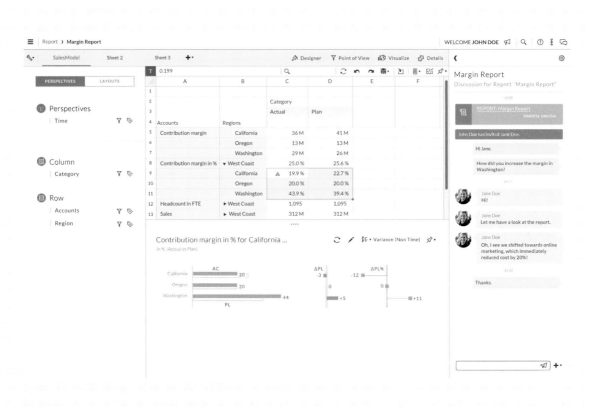

FIGURE 7.2
SAP Cloud for Planning allows collaborative financial planning and analysis sessions.

found. Business users can define their data model based on their business needs and not technical guidelines. They can easily insert data, set up new planning scenarios, and create their plans. There is no need for complex configurations such as preaggregations, normalizations, indices, or star-schemas. By selecting specific cells, the application instantly creates charts for the user to better understand the numbers. On the left side, users can further navigate, filter, and drill down the planning data. On the right side, users can collaborate with their colleagues via chat and discuss their planning results.

An integral part of SAP Cloud for Planning is social interaction and collaborative decision-making in the planning process. It is key is to enable open teamwork, such as reaching out to co-workers to get their opinion while staying in the context of the current task. When the collaboration pane gets activated it infers the current application context and links it directly to a conversation or calendar event. All participants can easily share and discuss their work with the integrated messenger. Even individual planning versions can be selectively shared to gather all relevant opinions before submitting them. The integrated notification center keeps all users informed about their work-related events such as Key Performance Indicator (KPI) alerts, new conversations, tasks and events. When users are not using the web client at that time, they receive email notifications instead. A built-in calendar is used to model all business events and implicitly works as a workflow engine.

Events can define approvers, reviewers, ask for data input and dashboards, and automatically trigger subsequent actions. All together, these components create a team workspace to organize, inform, and share all planning-related information, without the need to switch tools and lose context. SAP Cloud for Planning eliminates the laborious handling of spreadsheets and detached email threads.

Cloud Architecture

The underlying cloud infrastructure speeds up the deployment time for users. Software can directly be accessed via a simple web browser – there is no need for any installation process. As SAP also maintains the software for all of their customers, new versions can be rolled out quickly. In other words, there is only one current version of SAP Cloud for Planning that is the same for all clients, always up to date, and includes the latest features.

In addition, there are still more synergy effects which a customer's IT department would not be able to provide in that scale. SAP as a cloud provider offers their applications on hardware that is shared between different clients. Customers pay only for what they use. This, in turn, leads to lower Total Cost of Ownership (TCO) over the application lifetime and a quicker Return on Investment (ROI). And if clients need more resources such as CPUs or memory, the cloud platform supports flexible addition of these with just a few clicks.

Libraries for Planning and Analytics in the Cloud

SAP Cloud for Planning combines general cloud software advantages with the specific benefits of HANA.

HANA as a development platform comes with different libraries to build a comprehensive cloud solution. Libraries for web-based UIs, such as SAPUI5, eliminate the need for client software. Calculation libraries, such as the planning engine, bring standard planning and distribution functions to applications.

Furthermore, SAP Cloud for Planning uses HANA's capabilities to easily integrate external data and metadata. For example, SAP has built a bidirectional integration with the existing SAP Business Planning and Consolidation (BPC) system. There are also special data-integration libraries such as HANA Cloud Integration (HCI) for data services, offering the possibility for complex synchronization mechanisms with transformation and automatic mapping of data between the Cloud and the customer landscape. The existence and seamless integration of all these libraries show why HANA is a platform for business applications and more than just a fast database.

7.3 Celonis Process Mining – A New Level of Process Transparency

Business processes are the heart and soul of every large company. Whether in purchasing, sales, production, or Customer Relationship Management (CRM), processes often involve many different experts, employ various IT systems, and span across several organizations and locations. Keeping track of what is actually happening in a large enterprise is becoming an increasingly difficult and time-consuming task.

Due to human error, high complexity, varying interfaces, and changing environments, not all processes run as they are supposed to. Purchase orders, for instance, are not always issued in time and sometimes carry incorrect price data which leads to extra work and slows down the organization. Other examples are sales orders that are delivered late because of hold-ups within global supply chains, production orders that have to wait for material from prefabrication, and so on.

Inefficient and slow processes are key cost drivers in any business and directly affect corporate performance. Therefore, organizations have a strong need to obtain transparency in their important business processes and understand where and why these inefficiencies occur. However, obtaining this kind of information automatically is challenging as IT systems are primarily designed for everyday business operations and do not support complex process analyses. Extracting and manipulating data manually, on the other hand, is time-consuming and requires expert domain knowledge.

Celonis Process Mining is based on HANA offering a new approach to process analysis. It enables users to see how their processes are actually executed and where inefficiencies and deviations from the standards are occurring. This project has been conducted by Celonis in the context of SAP's Startup Focus Program.

⚆⚆ POINT OF VIEW

Process Transparency

An important customer using Celonis Process Mining on HANA has over 300,000 employees, a revenue of over €75 billion, and over 60 different SAP systems worldwide. Daily business involves hundreds of thousands of SAP documents, transactions, and activities, adding up to a huge volume of data. This company faced the challenge to assess and manage its data and processes in this complex environment. Instead of modeling processes on the drawing board, they needed to know how processes were really running in different parts of the organizations. This approach promises to spot potentials for process improvements that could not be detected before. In the end, the goal was to improve efficiency, save costs, and increase productivity. Celonis Process Mining helps this company and many others to achieve these goals in a unique new way.

⚐ SOLUTION

Celonis Process Mining

HANA, in combination with Celonis Process Mining, offers a solution to these challenges by providing powerful tools to continuously monitor processes. The idea of process mining is to reconstruct actual process flows from digital traces, i.e., table data, logs, and timestamps, which are created every time a user works with an IT system. Celonis Process Mining leverages the capability to access live data from SAP systems without the need to load, transform, or aggregate data. After mining this information, Celonis is able to visualize actual process graphs in a very flexible manner. This introduces a paradigm change from a priori knowledge (modeling the process the way one thinks it runs) to a posteriori insights (seeing the way the process actually runs).

Based on the reconstruction of actual process flows, Celonis Process Mining allows end users to perform complex process analyses such as:

Variant analysis

In real life, processes do not always follow the ideal path. For instance, a purchase order might be changed several times, linked to the wrong vendor, or delivered too late. Variant analysis is an easy-to-use tool for analyzing all of these different path variants of the process. Users can see the most frequently followed path (the happy path) and all potentially unwanted exceptions that are the root of unnecessary complexity (see Figure 7.3).

Root cause analysis

Leveraging the analytical capabilities of HANA, users can seamlessly drill down into and inspect all raw data objects which belong to the analyzed processes (e.g., customer master data, sales order header, position, schedule line data, and delivery documents). Users are enabled to find root causes for delays or deviations within the processes.

Throughput time analysis

Throughput times can be displayed and analyzed right within the interactive process graph (see Figure 7.4). Based on these, users can easily

spot the exact location and reasons for hold-ups and delays within the process.

Continuous process control

Users can easily build intuitive analysis screens which provide constant operational feedback about specific questions regarding processes.

Celonis Process Mining can be used for many different domains, such as purchasing, sales, logistics, production, accounting, master data maintenance, and customer service.

⚷ ENABLEMENT BY HANA

A Powerful Backbone with HANA

Process mining requires a data processing backbone that is able to handle large amounts of data. The technology needs to process all data associated with each process instance without using aggregates. As traditional databases read data row by row, their scan speed is not sufficient for mining this huge amount of information. In contrast, the columnar orientation, dictionary encoding, and parallelization of HANA enable interactive data mining and, thus, the required analysis.

Another important aspect of process mining is data access. Extracting and reintegrating data from various enterprise systems to an additional

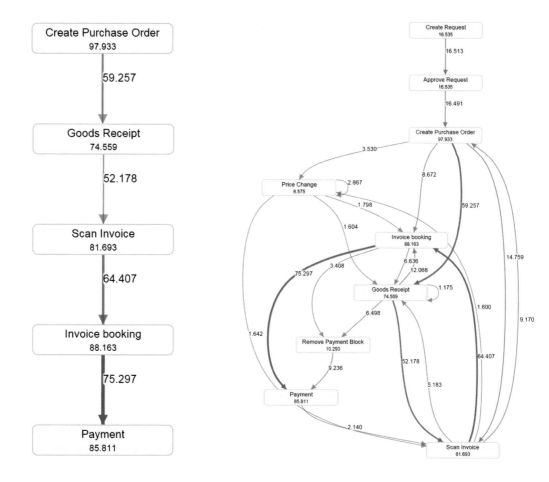

30% - View: The "happy path."
The most common process variant.

90% - View:
Most of the variants can be analyzed.

FIGURE 7.3
Variant analysis of a mined business process

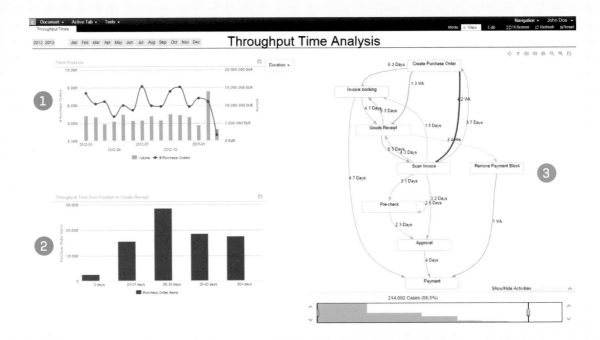

FIGURE 7.4

Example analysis of a production process: ❶ Flexible drill-downs to any attribute associated with the process (e.g., time, material, plant) are possible. ❷ The distribution of throughput times shows variations in the execution of a process. ❸ The Process Explorer displays throughput times directly in the process graph and allows the user to spot bottlenecks and hold-ups.

warehouse is a slow and costly process and increases complexity. If customers already run their Business Suite on HANA, Celonis Process Mining can directly work on the raw data. In case of data being distributed over multiple systems, the HANA platform provides easy access to the global SAP infrastructure. Therefore, the applications can be directly connected to HANA and updated continuously.

columnar orientation, dictionary encoding, and parallelization enable interactive data mining

7.4

Signal Detection in Big Marketplaces

Online shopping offers numerous conveniences: shopping anytime and anywhere, price comparison, user reviews, complementary goods, purchasing from foreign countries, and many more. While the front-end in e-commerce (business via the Internet) presents simplicity to its users, the back-end usually is a complex network of databases and servers. As e-commerce keeps growing at a rapid pace, this network will grow larger and become more complex, storing a huge amount of data. Nevertheless, e-commerce giants with over 100 million registered users are starting to realize that they have a valuable source of information within these networks. Not only can these companies leverage the collected data for new insights, they can also use it to reveal previously unknown patterns that can be used to improve their overall business processes.

With all this data, one concept that online marketplaces want to realize is signal detection. This concept concerns monitoring the operational health of the marketplace based on a set of statistical indicators. Any statistically significant deviations from expected behavior – as provided by sophisticated forecasting algorithms running over the huge set of data – should emit a signal indicating that something important needs attention. These signals come with varying severity; a less severe signal could indicate that the number of new users is lower than expected, while a high severity issue could correspond to fraudulent activity. Whichever is the case, it needs to be brought to the attention of analysts as soon as possible. Otherwise, major irregularities can cause a drop in customer satisfaction and result in a significant loss of revenue.

However, traditional systems for data storage and analytics involve the trade-off between the ability to consume massive datasets and the ability to detect changes in data quickly. Such technological limitations have made the idea of signal detection for the entire marketplace almost impossible to realize.

HANA, with its in-memory columnar storage and parallelization, enables extracting value from big datasets almost instantaneously. SAP leverages this to realize the idea of signal detection in a co-innovation project with one of the world's largest online marketplaces. Data scientists can now be directly informed about anomalous behavior in the marketplace and can act on resolving it without losing precious time.

POINT OF VIEW

Detect Anomalies as Soon as Possible

In big marketplaces, it is essential to detect or even predict anomalous behavior as soon as it occurs. The process of detecting and predicting such anomalies must be automated, cover the entire system landscape, and be real-time. Anomalies must be presented to the responsible employees in a simple and powerful interface that empowers them to take the required actions to resolve this behavior immediately.

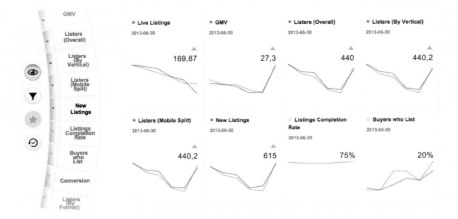

FIGURE 7.5
A snapshot of the dashboard landing page, showing key metrics with their associated signals

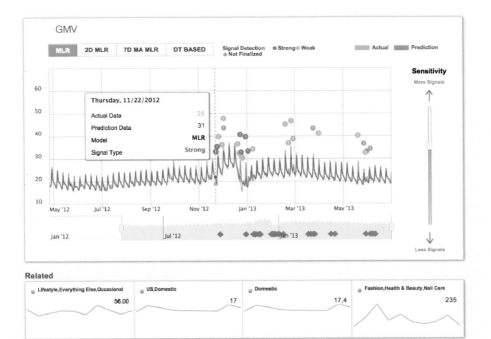

FIGURE 7.6
Signal detection in the GMV category

SOLUTION

Insights into Online Marketplaces

With HANA, the marketplace is now able to analyze multiple terabytes of data for signal detection in real time. The innovative solution identifies indicator metrics to be measured, runs multiple forecasting algorithms on the entire marketplace data to determine the best suited algorithm for each indicator, monitors over a million such metrics simultaneously, and detects statistically significant deviations between forecasted and actual data – all of this in a few seconds.

To get access to the monitored marketplace, the solution comes with a dashboard that comprises of all the high-priority signals applicable for that day across key categories (see Figure 7.5). By having a look at this dashboard, data scientists are able to quickly identify which areas need immediate attention.

Monitoring

With over 100 million users, the online marketplace generates billions of data entries every day, such as new listings, registrations, and orders. Even after preprocessing the data, traditional databases only provided computational power for analyzing a small fraction of the data in real time. This meant selecting only a few metrics to monitor while acknowledging that there could be severe issues in the unmonitored areas.

The co-innovative solution gets rid of this by using HANA's powerful computational engine, which can run analytics on all the data that the marketplace generates every day. In the background, algorithms run on all possible combinations of existing metrics, and thus, check over a million indicators for possible signals. The data scientists also have the flexibility to configure new metrics and run signal detection for these.

Signal Detection

An important prerequisite for signal detection is high accuracy for forecasting expected behavior. SAP's solution is able to run multiple algorithms to determine the best one for each metric, resulting in highly accurate forecasts. Furthermore, it speeds up the comparison of forecasted with actual indicators when bringing data from multiple sources into one place. Figure 7.6 shows an example of signals in the Gross Merchandise Value (GMV) category. While the graphs draw a comparison of actual (orange) and predicted (gray) data over time, colored dots highlight detected signals as strong (red) or weak (orange) differences. Signals are colorized to show severity depending on their statistical strength. In this example, there was a difference of about $3 million between the actual and predicted value for merchandise sold in November 2012. Such signals are usually detected on the fly, saving valuable time and money as soon as large differences occur.

Handling the Complexity of Big Data

HANA is well suited for Big Data challenges with its constant addition of analysis tools and seamless integration of statistical software such as SAS. In the following, two HANA features are highlighted that support the presented solution in particular.

Data integration

The data required for signal detection comes from different systems and in different data formats. HANA can interact with many special-purpose software solutions and offers data transformation tools such as SAP Landscape Transformation and SAP Data Services, which bring all data into one common format to store them in one place in memory. This reduces the time between data generation and consumption for users and applications.

Real-time analytics

It is important to detect signals as soon as they happen. A critical signal could translate to lost revenues of millions of dollars each day. As HANA eliminates the need for materialized aggregates, it becomes possible to analyze the entire marketplace on every desired level. It also empowers end users to create metrics on the fly and run analytical queries in real time. HANA enables this by using in-memory columnar storage, which facilitates quick scanning of individual columns. Combined with dictionary compression, the data footprint is reduced which thereby accelerates query execution times. Furthermore, HANA gets rid of slow disk reads, leverages multiple CPUs for the parallelization of tasks, and integrates analytical software allowing complex algorithms to be run on a huge amount of data in sub-seconds.

CHAPTER EIGHT

HIGHEST FLEXIBILITY BY FINEST GRANULARITY

With HANA, users are enabled to explore data by freely choosing the final selection and aggregation of raw data on its finest granularity. Users can search through data by successively adding or removing selection criteria for any attribute. They can also start with an aggregation of their choice and then drill down to its respective line items – all in real time. Any traditional database can scan, search, and aggregate stored data. Yet, if users have to wait for minutes or even for hours until each step is finished, it effectively means that there are no capabilities for scanning, searching, and aggregating.

Outline of this Chapter

SECTION 8.1 **ENTERPRISE SIMULATION** What-if analyses are a well-established approach to understand the potentials and inefficiencies of a company. With the HPI Enterprise Simulator, business analysts can perform such analyses on actual enterprise data in real time.

SECTION 8.2 **TAKING THE SPEED CHALLENGE** Flexibility is a key requirement for a large Consumer Packaged Good (CPG) company whose main table contains over 200 billion rows. As a replacement for their data warehouse, the company introduced a HANA-based solution which answers user-specified queries on live data for thousands of users simultaneously.

SECTION 8.3 **FRAUD MANAGEMENT** On average, 5% of an organization's revenue is lost to fraud. SAP Fraud Management is a cross-industry application that addresses this problem. It analyzes, detects, investigates, and prevents fraud and irregularities in high-volume data environments.

8.1 Enterprise Simulation

Companies invest a significant amount of time in the yearly budgeting process and resulting quarterly or monthly forecasts. This process is often seen by management as inefficient given the volatility of markets and enterprise structures. In this context, the what-if analysis has been established with the goal to closely model cause and effect in an enterprise and its environment. This functionality can be used for the budgeting process or as part of a forecast for scenario evaluation in terms of their goal fulfillment. However, while this theoretical model has well-defined semantics, it still lacks proper tool support.

The HPI, in cooperation with the SAP Innovation Center Potsdam, has connected with a Fortune 500 company in the consumer goods industry in order to discover their needs for enterprise simulation and create a new simulation tool.

POINT OF VIEW

Key Requirements for a New Simulation Tool

Together with the Fortune 500 company, the following key requirements for the simulation tool were identified:

Flexibility   Adaptability to new use cases without additional programming efforts
Interactivity   Sufficient performance for interactive decision-making during planning runs
Collaboration Multi-user support for collaborative development of joint simulation scenarios

---

SOLUTION

Real-Time Hierarchical Calculations

These identified requirements can be addressed effectively with HANA's capability for direct execution of analytical queries on transaction data. To meet the challenge and leverage the power of HANA, the HPI has developed the HPI Business Simulator – a tool that allows companies to define and calculate what-if analyses in seconds. The main purpose of this tool is to enable companies to flexibly simulate scenarios in real time. Users can easily configure and perform their simulations without the development overhead of custom-built simulations. This serves as an enabler for ad-hoc decision support, planning, and forecasting – positively impacting multiple areas within a company:

Purchasing

Material costs can be simulated in dependencies of commodity prices and currency fluctuations.

Production

Costs can be simulated based on production paths, machine allocations, transportation costs, rejection rates, energy consumption, energy prices, and more.

Sales

The sales volume can be simulated using drivers such as unit price, and economic factors, including buying power and competitive vendors.

Controlling

The value drivers can also be consolidated and used for profitability simulation. On the management level, executives gain more process transparency through real-time information access and can see the impacts of strategic decisions.

IT infrastructure

Data redundancy is eliminated, resulting in less need for storage and reconciliation processes. With this, data inconsistencies can be eliminated as well.

The HPI Business Simulator builds on the concept of value driver models [GN00]. Value driver trees, such as the DuPont model, are well-known methodologies to model Key Performance Indicators (KPIs) with independent linear equations or – in the case of input-output structures – with systems of linear equations. Using value driver models, activities and decisions in a company are focused on the core factors that drive the KPIs, e.g., the operating profit. Furthermore, their usage increases collaboration across departments, leading to more aligned operations. This results in reduced effort required for planning and the development of more realistic plans, as the KPIs are directly connected to the operational drivers. Drivers can influence multiple other drivers, e.g., an increased sales volume influences both net sales and variable costs.

A value driver model is a directed graph consisting of a set of nodes and their connecting edges. The nodes can either represent a KPI or a value driver, which are hierarchically modeled using the edges. Each node is a value driver that either represents a data source or is calculated based on other value drivers. Each parent is a combination of its children, which can either be connected by basic mathematical operators or more complex systems of linear equations. Figure 8.1 shows a simple value driver model for the operating profit. Value drivers

Businesses can create and run custom what-if analyses in seconds, enabling ad-hoc decision support.

(nodes) are connected by mathematical operators that determine how drivers influence each other. Value driver models can be filtered along the product, customer, location, and time dimensions.

Based on this concept, the HPI Business Simulator consists of three major parts: the value driver model configuration, the parameterized simulation execution engine, and the calculation engine.

Flexible Configuration of the Simulation Model

The simulation model is usually configured by domain experts who define the relevant value drivers and their dependencies. Each node in the value driver model has a name, a definition of dependencies to other nodes, a unit, the dimension it supports, and possibly a link to the data source it represents. After the model has been defined, the IT department can create the data binding for each value driver. The data binding includes the definition of the value that has to be aggregated, the attributes that define the dimensions a simulation supports, and additional information that specifies the value driver.

As an example, for the value driver sales volume, the attribute quantity would be the value that has to be aggregated. Customer, location, product, and time would be attributes that specify the supported dimensions, and these would ensure that only relevant values for the sales volume are aggregated. The HPI Business Simulator supports the creation of multiple models that can be stored and reused for different scenarios.

Real-Time Simulation

Once the value driver model has been configured, simulations are initiated by adjusting (overriding) values with the nodes of the driver model. The simulated impact is visualized in relation to the actuals, plans, and forecasts. Users can set filters on the dimensions, e.g., by customer, location, or product, to explore the impact of the simulation run in detail.

The prototype application depicted in Figure 8.2 calculates a P&L statement. This calculation can be directly executed on company transaction data that can exceed 300 million records – without any prematerialized aggregates. This enables users to adjust simulation parameters such as the inflation rate of a country on any level in the value driver model and on any filter dimension. In the example shown in Figure 8.2, the Sales Volume node is selected and we can see that the sales volume for the month of August has increased by 12.3%, with a direct impact on the gross sales and the Cost of Goods Sold (COGS). The HPI Business Simulator allows users to immediately see the impact of changes on all levels of the P&L statement.

Complex Calculations

Several nodes in the value driver model cannot be calculated simply by aggregating a single dimension, but instead require complex input-output calculations. As an example, the calculation of product costs is a complex task due to the highly

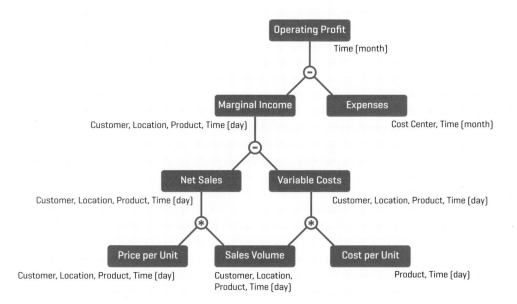

FIGURE 8.1
A simplified value driver model for the operating profit in the form of a value driver tree

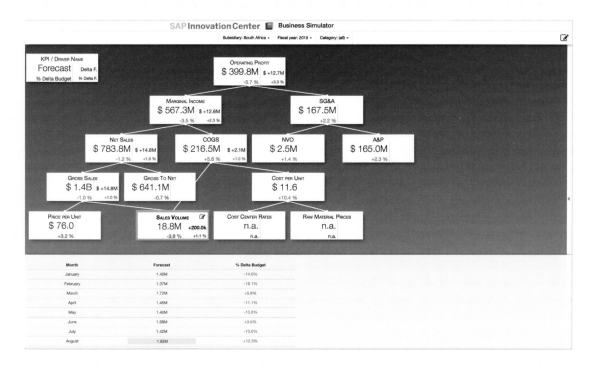

FIGURE 8.2
Drill down in the Profit & Loss (P&L) simulation

distributed types of cost components. Those costs incorporate material prices, manufacturing costs, labor rates, electricity rates, machinery costs, depreciation, and everything else required during production. These cost components have to be mapped based on the bill of material and the bill of operations, which is also referred to as routing, in order to be part of the overall costs of a product. During the calculation process, all components are mapped to a system of linear equations that is solved by the tool to produce the simulated costs.

queries that frequently occur during a typical simulation session. This can reduce query execution times and decrease the overall system load at the same time. Complex calculations such as necessary for product costing can be highly parallelized in HANA by leveraging multi-core CPUs. Thanks to the power of the HPI Business Simulator, businesses will be able to create and run custom what-if analyses in seconds that can enable ad-hoc decision support, planning, and forecasting.

 ENABLEMENT BY HANA

Aggregation on the Transaction Level

A key enabler for this approach to enterprise simulation is the ability of HANA to flexibly aggregate data with sub-second response times. With the simplified data model, it now becomes feasible to directly use the transaction data for enterprise simulations. The simulation calculations can also use external data such as commodity price development, raw material cost development, and labor cost variations. The columnar data layout of HANA leads to another benefit – as the mapping of the simulation model to a database table accesses only the columns required for the calculation, the amount of processed data is minimized.

Additionally, HANA's transparent aggregate cache further accelerates recurring aggregate

8.2

Taking the Speed Challenge

With the rise of Big Data, a large Consumer Packaged Good (CPG) company realized that they could extract more value from their data. However, their traditional data warehouse still reflected the hardware limitations of the late nineties, when storage was relatively expensive and response times were limited by access time necessary for retrieving the data from disk. Even if such a warehouse is able to answer strategic questions – e.g., "Which products contributed to the differences in sales figures between regions in the past 14 days?" or, "What factors contributed to the increase in shipping costs compared to the previous month?" – it does not meet the flexibility which is required these days. Specifically due to a fixed star schema, the traditional data warehouse is not able to provide fast answers to queries which do not fit a predefined template with strict parameterization.

For every company that wants to become a world-leading company, it is insufficient to be able to answer only questions that change little over time – they need to analyze whatever their business demands.

SAP's response to the company was a tool based on HANA with a much broader range of applicability than the traditional data warehouse. Not only is the goal of this tool to provide information for a handful of strategic decision makers who would chart the company's path forward, it would also open the data silos and answer questions for a much larger array of operational staff aiming to resolve daily issues. Thus, the company puts information at the fingertips of every employee, not only the boardroom dwellers. This unlocks data-driven innovation potential across the entire corporation. The company is able to respond in real time to challenges as they arise instead of trying to guess ahead of time what those challenges could be.

POINT OF VIEW

A Quarter of a Billion Transactions Every Day

The company had a number of requirements. First, as it is impossible to predict the exact questions which will be asked by users, all data must be available at the highest level of granularity. No materialized aggregates can be used as these limit the expressiveness of questions which can be answered without long response times. Second, the data basis should comprise at least the information of the past three years. Third, the company needs to be able to work with up-to-date information – updating only once a day does not provide the desired flexibility. Fourth, the back-end must support thousands of concurrent users, rather than the three to five users that traditional data warehouses are accustomed to. Finally, the system has to be responsive – no query should take longer than a few seconds to run even if the system is under heavy load. With only a few thousand transactions, these requirements could also be handled by a traditional database. However, this company records over a quarter of a billion transactions every single day.

☐ SOLUTION

Eight Seconds

A traditional data warehouse could not deal with this Big Data challenge. For this reason, the company chose HANA in order to meet its requirements. The queries the company wanted to test ranged from queries spanning across years to others looking only at a restricted subset of data for a single month with aggregation levels of an hour or less. The query types, their acceptable response times, and their frequency are shown in Figure 8.3. The eight second upper limit is based on studies which indicate that this is the maximal acceptable response time for almost 80% of people [Shn84].

With these requirements in mind, SAP deployed the most recent version of HANA onto a 16+1 node cluster – each node in the cluster contained eight 15-core sockets and 2 terabytes of RAM. Such a large cluster was necessary because the main table of the company contained over 200 billion rows – a challenge HANA handles by partitioning tables and distributing them over the cluster nodes. After this step, HANA loaded the data into the cluster and was able to insert new data every hour at a rate of three million rows per minute. This rate exceeds the maximum required insert rate, even for peak periods. This shortens the delay from the moment the data is created until the moment it can be turned into usable information for the business.

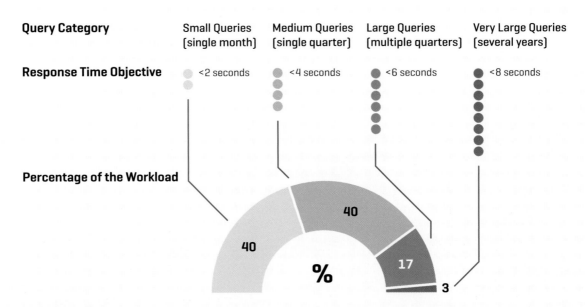

FIGURE 8.3
Distribution of queries in a typical company's workload and the longest acceptable response times

Based on the deployed system, SAP ran multiple tests for each kind of query with 1,000 active users and a single user as the baseline. The test assumed that each active user would execute roughly between 20 and 40 queries. For each test, three quantities were calculated: shortest, longest, and average response time. The test results are shown in Figure 8.4.

With these results, the company decided that HANA should provide the new back-end. The system not only performed very well for single users, but also when handling a high number of active users. The average response time remained below the target for every kind of query, even very large queries dealing with several years of data.

Only in rare cases was the eight second bound slightly exceeded because of a single, very large query.

With HANA, the company was able to achieve the vision, and, perhaps, something more. One time-tested axiom of business is "knowing the customer." Small businesses have always been able to honor this dictum by paying attention to the habits of their core customers. Multinational corporations, on the other hand, had no way to even know who their core customers were (and much less their habits) because the data they analyzed was not detailed enough. However, the company's insistence on making data available down to the item level now finally gives them the opportunity

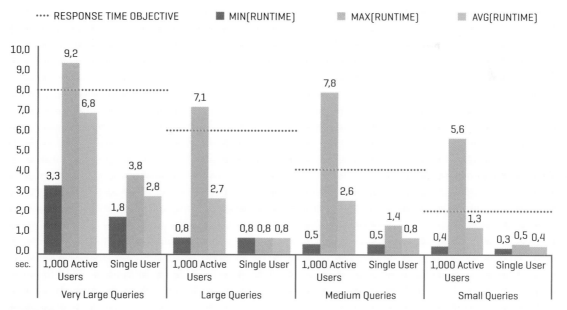

FIGURE 8.4
Benchmark outcomes for reporting queries: response time is below eight seconds on average

to disrupt their competition by making it possible to analyze each of its millions of customers. Now, they can determine who their core customers are, what their habits are, and whether particular promotions are likely to influence them.

Comparison with a Traditional Data Warehouse

The initial implementation was based on a traditional database. However, the traditional back-end did not perform – simple queries took a very long time to produce a result, assuming a result was produced at all. Thus, to provide a responsive system, the data warehouse was forced to preaggregate. Such materialized aggregates, however, made it much more difficult to update the system. For this reason, updates could only be done once a day (usually in the middle of the night) in order to compute the aggregates. The situation became even more difficult when the dimensions evolved over time instead of remaining static. Every evolving dimension invalidates the materialized aggregates and forces a computation-expensive recomputation of all aggregates in the system.

As long as the data was limited to a subset of the yearly data and the number of concurrent users was in the range of three to five, the system could return results for queries in a reasonable timeframe of five to ten seconds. However, if the number of users grew beyond ten, the system became unresponsive. Instead of the scalable, flexible system with a straightforward data model that the company had envisioned, with a traditional database they had received a high-maintenance, brittle system with a complicated data model that could not scale.

HANA, in comparison to a traditional database, does not posses any of those downsides and instead, enables the highest possible flexibility at the finest level of granularity – without relying on any predefined aggregates. With its combination of in-memory technology and columnar storage, required aggregates can be built on demand, thus HANA allows companies to ask any desired questions at any time.

In summary, HANA enables businesses to draw new insights from vast quantities of fine-granular data with very short response times.

> Companies can determine their core customers, their habits, and the influence of promotions.

8.3

Fraud Management

Organizations lose an estimated 5% of their annual revenues to fraud [Ass14]. This translates to a potential projected global fraud loss of nearly $3.7 trillion in 2014. At the same time, organizations have to comply with anti-money laundering laws and others such as the Bribery Act and the Foreign Corrupt Practices Act.

Companies have solutions in place to detect fraud and ensure compliance. However, only 1% of all discovered fraud is detected by IT controls and many of these solutions have limitations that these companies would like to overcome. The main problem in automated fraud detection is the accuracy of algorithms. In order to detect real fraud cases, current systems have to accept a high number of false positives – that is, mistaken fraud warnings. False positives cause much effort without creating any value. In some cases, they even result in reputation damage if payments were stopped due to a false alarm. False positives are often a result of inflexible detection rules and complex and time-consuming procedures to update them. Usually, this goes along with limited calibration and simulation capabilities which, at best, can be done in long batch runs over the weekend. During the period that the detection rules are not adjusted to the continuously changing fraud patterns, more false positives are produced. At the same time, there is no learning curve from processed alerts and companies cannot discover insights that are actually hidden in their data. On top of these limitations, the existence of data silos prevents the detection of even simple fraud cases as the data is stored in separated business applications. Consequently, companies very often detect fraud too late and at a point in time when the money has already left the organization. Realizing the legal claims at this point is often costly or practically impossible.

POINT OF VIEW

Preventing Fraud

Companies would like to improve fraud-detection with a unified toolchain that detects fraud as it occurs, while also enabling company employees to efficiently resolve such potential frauds. For this, several technical challenges have to be overcome: (1) the huge amount of data that needs to be processed, (2) the versatile sources, natures, and formats of the relevant data, and (3) the need for fast detection of fraud as it occurs.

SAP Fraud Management

SAP Fraud Management is a cross-industry application to analyze, detect, and investigate irregularities and thus prevent fraud in Big Data environments. It covers both fraud and compliance scenarios that can be based on SAP and non-SAP data sources.

The solution targets ERP users, and comes with predefined content for detecting fraud in the areas of Finance (procurement, sales) and Human Resources (travel expenses). On top of this, the solution supports a wide range of industry-specific use cases, very often with content provided by partners of SAP:

> fraudulent insurance claims
> money laundering in banking
> tax-related fraud in the public sector
> energy theft in utilities
> hacked SIM cards in telecommunications

One of the key aims of SAP Fraud Management is empowering the business users. The solution comes with intuitive user interfaces to support the entire cycle of detection, investigation, and prevention – all in one solution.

The most relevant information, such as Key Performance Indicators (KPIs) and various statuses, is pushed to a dashboard which can be easily and individually configured by the end user. From there, the user can navigate to the most used functions with just a single click and without navigating through multiple menus (see Figure 8.5).

Within SAP Fraud Management, users can create and change detection strategies without any support by the IT department, just by customizing and parameterizing the predefined rules as they want. If this is not enough, they can use HANA to create new rules without programming knowledge – simply by using Decision Tables. Decision Tables extract the core decision logic in a tabular structure that is quick to read, clean, and understandable. With this approach, the power is shifted to people that understand the business and are close to the fast-changing fraud patterns. The business user can run on-the-fly simulations and calibrations in order to reduce false positives without losing correct fraud cases. The analysis of the patterns is graphically supported with Sankey diagrams (see Figure 8.6) and tree charts.

For further investigation, the users are presented with specific and relevant information in order to allow them to make decisions or drive further actions and collaborations across departments. For instance, consider the investigation of a car insurance claim. The user can see all information about the policy, the claims, and claimants. The user also can see detailed information about the accident itself, including pictures of the damaged car and geographical information. On top of this, there are network analysis capabilities to detect hidden connections such as relations between multiple claims of a policy holder or even closed loops where the same people are involved in multiple claims in the same or similar roles.

FIGURE 8.5
Dashboard of SAP Fraud Management

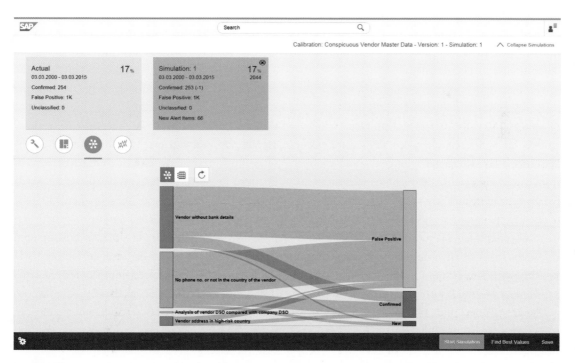

FIGURE 8.6
Simulation analysis of fraud management with Sankey diagrams

⚿ ENABLEMENT BY HANA

Business Innovation Powered by HANA

The power of HANA opens up new capabilities in fraud detection and partner screening.

Fast screening performance

Companies across all industry sectors have to deal with more and more complex data. Many fraud cases, even simple ones, cannot detected because screening on a fine-granular level in Big Data environments would take weeks. With the power of HANA, SAP Fraud Management can identify complex fraud patterns with multiple detection rules against tens of millions of records in just a few seconds. Use cases in healthcare, telecommunications, or banking with billions of records become feasible.

Interactive, on-the-fly calibration and simulation to optimize detection rules

Due to the high screening speed, the nature of the detection process changes completely and evolves from batch to interactive. Simulations with different parameterization can be done on the fly. The end users become empowered to optimize the rules themselves, without IT support, which leads to a continuous calibration of the detection rules. This is essential to reduce the number of false positives and keep up to date with fast-changing fraud patterns.

Predictive capabilities to detect unknown patterns

The remaining step for developing a self learning-system, which can discover closed loops, is to combine on-the-fly simulations with predictive analytics. With the embedded predictive algorithms, the system can detect outliers, improve the parameterization, and detect unexpected fraud patterns. Not only data mining experts, but also business users can use these classification algorithms. The system maintains the estimated costs of undetected fraud cases versus the cost of processing false positives. Using this, it generates optimized rules with optimized parameterizations.

Fuzzy search in unstructured data

With fuzzy search, users can search for any kind of terms in structured and unstructured data. This is essential for both investigation and detection. Investigations show that fraud and bribery is very often not done only once, but rather repeatedly by the involved individuals after they devised a seemingly safe fraud strategy. Although the damage of a single fraud case might be low, the real value of an investigation comes into play once the structural problem is detected. With fuzzy search, the investigators can, for example, easily relate the current case with any other payment or financial item. Very often, people use code words to communicate across departments or even companies. With HANA's built-in search, the investigators can search for terms in unstructured data, such as: text fields in General Ledger (G/L) line items, scanned invoices from vendors, and any documents stored in the database. Even findings from previous cases might be of interest and can be related to a current case. Another important use case for fuzzy search is the screening of partners and addresses. In their own interest and for legal reasons, companies need to know with whom they are doing business. Therefore, companies must establish screening procedures and compare names and addresses against lists offered by data providers (e.g., sanction lists or politically exposed person lists). In this context, it is crucial to identify slightly differently written names or addresses as one and the same. This fuzzy search must be fast as it is often used in high data-volume environments. A comparison between HANA and a traditional database shows that the screening time for a company with around 30 million customers can be reduced from one week down to three hours.

Multiple data sources

Very often, only the combination of data from different sources reveals obvious fraud cases. HANA can handle data from different sources such as data from purchasing, inventory, and accounting. Only if these sources can be combined, patterns like overpayments or payments without delivery can be detected.

Real-time integration into SAP and non-SAP business processes

As soon as companies become real-time enterprises and steps in procure-to-pay and order-to-cash processes are fully automated, real-time integration of fraud management into SAP and non-SAP processes becomes essential in order to stop a process if appropriate. SAP Fraud Management can, for example, run integrated into SAP's payment program and block a payment until an investigation results in a decision. In online stores, it is also crucial to screen partner data against sanction lists during a transaction without delaying it.

For the future, we plan to extend the SAP Fraud Management solution to cover additional purposes such as process controls, fraud detection, partner screening, as well as internal and external audits.

CHAPTER NINE

EXPLOITING THE WINDOW OF OPPORTUNITY

T he short time period during which an otherwise unattainable opportunity exists is known as the window of opportunity. To seize these opportunities, it is essential to have access to all information stored in an enterprise system as soon as possible. With HANA delivering information in real time and SAP Fiori able to run on any mobile device, information is available anywhere and anytime.

Outline of this Chapter

SECTION 9.1 **OMNI-CHANNEL RETAILING** A luxury fashion brand wants to provide a personalized shopping experience for each of its customers. A mobile application provides sales associates with an assembled profile of each customer that combines their past shopping behavior with information on the customer's fashion related social network activities.

SECTION 9.2 **CONAGRA – FROM INSIGHTS TO ACTIONS WITH TOTAL MARGIN MANAGEMENT** The largest North American private-brand packaged food company wants to have a detailed understanding of when, where, and how costs are incurred and predict the impact of business changes on these costs. The presented Total Margin Management (TMM) tool will give ConAgra managers the ability to approach their operational data and make informed decisions with the guidance of enhanced forecasting capabilities, and by doing so, exploit their windows of opportunity.

SECTION 9.3 **RISK ANALYSIS DURING NATURAL DISASTERS** Long before the outskirts of a hurricane reach the coast, the work of reinsurance analysts begins. The Hurricane Damage Prediction application allows analysts to predict the extent of the imminent damage, therefore bracing the reinsurance company for the upcoming losses and enabling it to take the appropriate actions.

9.1
Omni-Channel
Retailing

Retail channels recently expanded to include more than simply retail shops. Nowadays, they also include online stores, websites, mobile apps, social networks, television, radio, catalogs, etc. This expansion has complicated the consistent representations of products, promotions, and prices to customers. Additionally, it has also become more complicated to understand a single customer holistically across all channels. Observing and understanding how to use these channels enables retailers to exploit the windows of opportunity by providing seamless consumer experiences. This ambition of today's retailers is typically hindered by several challenges:

› keeping data updated across all channels
› using the appropriate data models to capture the versatile data structures from different sources
› enabling seamless coordination between IT, marketing, and sales departments
› applying thorough analysis of the user-generated data (in particular feedback, reviews, and preferences)
› providing a seamless User Experience (UX) through simple and responsive User Interfaces (UIs)

Such problems require close collaboration between domain experts and software engineers. In order to jointly design and develop an application running on HANA to solve these problems, SAP followed their co-innovation model with a major luxury fashion brand.

 POINT OF VIEW

Seamless Consumer Experience

Today's retailers have been competing to provide a seamless, personalized, and unique consumer experiences across all of their shopping channels. This seamless experience ensures a consistent representation of information across all channels. Retailers want to track all channels in order to construct a holistic view on their consumers, which will enable the retailers to exploit their window of opportunity.

SOLUTION

Online and Retail Stores Blur

Nowadays, retailers rely on technology to support them in delivering superior consumer engagement, and enable them to better understand their consumers' needs, and thus, serve those needs accordingly. Such engagement yields more consumer satisfaction and results in higher consumer loyalty for these retailers. Towards this end, the co-innovation partner collaborated with SAP to develop a mobile application which provides their sales associates with real-time information about consumer preferences and the possibility to show additional product information to the customer.

This application gathers the shopping preferences, experiences, and wishes of the customers. Additionally, it monitors the specific brand-related publicly available activities of consumers over social networks. This data is combined with in-store data to construct an assembled profile of the customer showing a holistic representation. Based on the customer profiles, sales associates can offer each customer personalized shopping experiences by presenting a selection of products based on their shopping history and interests. Moreover, sales associates can suggest accessories or alternative garments to their customers. These are based on the insights generated through predictive analysis of customers' shopping history and text analysis of their social network activities – such as recognizing

positive or negative statements by scanning for keywords, e.g., in a public comment on a product shared on a social network. The corresponding data is stored in HANA and analyzed in real time to provide the necessary information for a particular customer visiting the store on a sales associate's tablet. In the coming example, we will follow our fictional customer Natasha Taylor in order to outline the benefits of omni-channel retailing.

While Natasha is attending a conference abroad, she takes a shopping break to visit the nearest store of our co-innovation partner. As the sales associates are empowered with the developed mobile application (see Figure 9.1), they can surprise Natasha with a unique, personal shopping experience. The sales associate can use the tablet application to make recommendations using predictive analytics fueled by Natasha's buying history, social network activity, and fashion industry trends. Natasha can examine her favorite raincoats which appear on the sales associate's tablet and can even watch videos of the products. If she does not want to buy her desired raincoat while abroad, the sales associate can seamlessly use the application to determine if these coats are also available in a store near Natasha's hometown. In the latter case, the sales associate can arrange for the clothes to be shipped to her home or reserve them for in-store pickup.

According to the co-innovation partner, this tool has given in-store customers access to the rich levels of immediate information that they have grown to expect from online experiences. Their CTO summarizes it thus: "Walking through the doors is just like walking into our website."

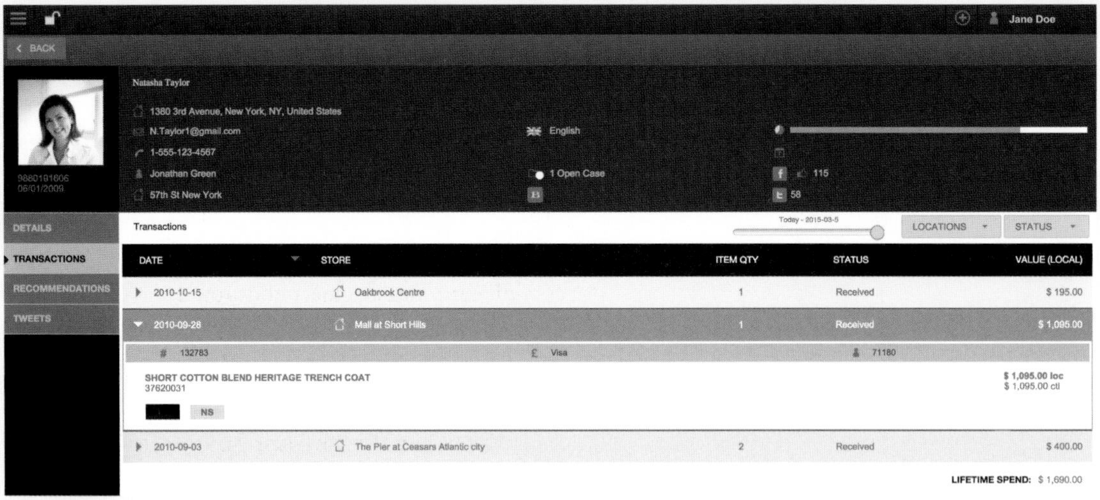

FIGURE 9.1
The mobile application: the profile of the fictional customer Natasha Taylor, showing her personal data and previous interactions with the co-innovation partner

⊷ ENABLEMENT BY HANA

Simple and Flexible Models – The Door to Omni-Channel Business

One of the main challenges for omni-channel retailing is capturing the versatility of data sources and structures, as every channel has its own requirements and semantics. With HANA, all incoming data is processed using simple data models and kept in memory. We have designed the data model of the application following the aggregate-free principle of HANA. This principle enables each channel to operate on its own views of the data. Additional views are added as new channels open or existing channels change. All data from every channel is accessible down to single user transactions. This approach enables easy and smooth coordination between involved departments. The speed of HANA ensures real-time processing of all channel-generated data, and therefore, keeps all data up-to-date. As all channels contribute to the same data, it becomes possible to realize a seamless UX. The application enables sales associates to seamlessly arrange the ordering and shipping of Natasha's favorite coats from an overseas store right to her house door.

The additional features of HANA, such as text analysis and predictive analytics, empower sales associates to provide high quality recommendations to guide the customers in their shopping experience; the text processing capabilities make it possible to integrate unstructured data from social networks with the highly structured data arising from traditional sales transactions, thus increasing confidence in the proposals. Shopping on-site finally offers the same advantages that were previously restricted to online customers.

The responsiveness of this mobile application is only possible because HANA is a full-stack platform and allows us to execute the business logic close to the data.

9.2

ConAgra – From Insights to Actions with Total Margin Management

The highly competitive environment of the food manufacturing industry puts pressure on companies to make increasingly accurate forecasts of demand in order to understand current and future sources of cost while maximizing profit. However, the large amounts of data produced in all steps of manufacturing pose a challenge to businesses seeking deeper insights to optimize current business processes and react to future needs.

Private-label food manufacturers following the low-margin, high-volume model must take on the task of keeping their prices affordable while producing vast amounts of products to sustain profitability. To compensate for the volatility of food and fuel prices, businesses want to have a detailed understanding of when, where, and how costs are incurred in order to manage product stock and minimize logistics costs. Combined with a forecasting ability on the same fine level of detail, production and brand managers can model the impact of business changes on their margin and profitability before executing these changes.

Comparison between different business scenarios is vital for effectively assessing the best course of action to continue increasing profitability and expanding the customer base. For instance, the logistics finance branch of a company, focused on transportation and warehousing costs, aims to be able to identify opportunities to minimize waste through unnecessary handling. It is essential to provide explanations for unexpected cost variances to enable action to be taken and for the model to be improved. Logistics finance requires the flexibility to model alternative distribution methods and rapidly assess the impact of fuel price changes so that they can reduce transportation costs. Many branches of business would benefit from the ability to predict such impacts, leading to increased profitability for the company as a whole.

ConAgra, the largest North American private-brand packaged food business, desired a solution that would allow them to not only gain deeper insights into their existing business model, but also accurately predict and shape future outcomes. SAP, together with ConAgra Foods as its co-innovation partner, resolved this challenge.

◯◯ POINT OF VIEW

Forecasts of the Future Enabled by the Understandings of Today

ConAgra wanted to enable its employees from all business areas to understand cost sources and draw insights within a unified framework. Different user roles, such as procurement finance or logistics, would be allowed to search through role-specific data and easily identify opportunities for current optimizations and future actions in their areas. For this to be possible, the solution must efficiently manage the huge amounts of data produced in the manufacturing process yet still allow managers to visualize this data at its base level in order to enable enhanced understanding of the product-customer profitability of the brand.

💡 SOLUTION

Unification of Perspectives for ConAgra

What ConAgra needed was a single system that could support all functional areas of business in understanding their cost sources at the finest granularity, with forecasting capabilities displaying modeled data on the line-item level based on insightful predictions (such as material or freight cost forecasts). Such a system would allow the company to both enhance existing processes and make accurate predictions to trigger future action by enabling the unification of broad cost insights with a deep understanding of actual and modeled items.

SAP's answer to this industry challenge is a tool called Total Margin Management (TMM), which will give ConAgra managers not only the ability to approach their operational data and effectively visualize the breakdown of their cost sources, but also make informed decisions for the future with the guidance of enhanced forecasting and modeling capabilities. At the time of this writing, the SAP-enabled solution at ConAgra Foods is being validated by its business users with a planned technical go-live in April and a business go-live in the summer of 2015.

Total Delivered Cost Analytics (TDCA) provides the first component of the TMM solution. TDCA lets managers separate costs into their base sources and visualize these, allowing greater understandings of each identified cost origin, thus enabling informed, responsive decisions.

The second component, Margin Management and Analytics (MMA), provides the ability to then take this broad cost decomposition and run analytics on the line items themselves to model future costs and revenues; this ultimately allows for the modeling of profitability at the product/customer level. Together, TDCA and MMA in the unified TMM tool will provide managers with the ability to identify and exploit current and future windows of opportunity.

TDCA is composed of two modules – a forecasting module that models all aspects of manufacturing costs, and an actuals module which breaks down data from a material ledger to provide a full cost structure for actual production that has taken place. The ability to separate and identify origins of costs from many types of operational data, as shown in Figure 9.2, will allow ConAgra to effectively align manufacturing and purchasing activities to meet forecasted needs, optimize

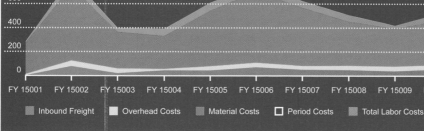

Americano Pizza Knutsford		FY 15001	FY 15002	FY 15003	FY 15004	FY 15005	FY 15006	FY 15007	FY 15008	FY 15009	FY 15010	FY 15011	FY 15012
Inbound Freight	Cost	3.20	80.00	21.92	45.76	45.28	72.00	52.16	48.96	39.68	55.36	66.08	43.20
Overhead Costs	Cost	9.12	45.59	38.12	11.55	34.25	37.47	30.89	35.81	39.34	36.87	11.65	19.59
Material Costs	Cost	256.82	617.68	303.90	271.78	462.03	592.05	484.88	367.21	315.95	371.54	520.71	424.04
Period Costs	Cost												
Total Labor Costs	Cost	14.94	74.69	34.96	62.03	67.93	52.02	55.83	55.39	25.28	60.09	65.39	68.27

FIGURE 9.2
View of a detailed breakdown of cost sources in a unified interface for enhanced planning (enabled by TDCA) and further analysis (with MMA) in the TMM tool

margin and profitability, and drive future demand. ConAgra employees will be able to adjust plans, understand their impact, and then decide whether to commit or discard these changes for the creation of multiple scenarios. The results of such activities are shown with an extension to the Enterprise Performance Management (EPM) Add-in, which provides an easy-to-use spreadsheet interface familiar to planners and analysts, eliminating the need to learn new reporting and analysis tools. All standard Excel functions can be used for both presentation and local calculations.

In addition to the rich spreadsheet-based planning experience, the TMM solution provides a comprehensive mobile-friendly web user experience. This enables several persons to create content and collaborate with each other, ranging from power users, defining planning models, to IT administrators, performing day-to-day operations and security audits, to end users, creating their own analytics.

MMA provides the ability to model and forecast Profit and Loss (P&L) at both the customer and the product level. Using MMA, managers are able to integrate actual results with the identification of business drivers and their conversion to forecast levers to steer the business. By looking at risk from a margin perspective, MMA will let ConAgra managers control projected costs and revenues based on the profit margin they imply. This approach does not isolate costs from revenues; instead, it addresses costs and revenues together as one.

Total Margin Management will provide managers with the ability to identify and exploit current and future windows of opportunity.

When TDCA is used in conjunction with MMA, as in TMM, a range of new business opportunities opens up. The increased efficiency in understanding and visualizing existing cost bases will allow ConAgra to meet changing business needs in a flexible manner. Beyond this, TMM also presents a scalable framework for business decisions concerning responsive action based on predictive analysis as well as adjusting their future direction from anticipated demand. ConAgra employees will not only be able to gain insight into what and why things happened in the past, but also see and understand things as they happen – and combine these insights with future knowledge of trends before they happen. The enhanced forecasting ability will enable these business users to leverage the power of enriched data to shape predicted outcomes.

ConAgra logistics experts, thanks to early insight into impacts from fuel prices, will be able to amend their mix of transportation modes to minimize the impact on profitability and maximize business success. SAP will enable

ConAgra to capitalize on opportunities as or before they arise by providing ConAgra managers with a deeper understanding of market trends and business-consumer relationships coupled with an enriched forecasting ability.

The power to draw insights from large amounts of data in real time in order to not only predict, but also shape future outcomes, is a game-changing advantage to companies in any domain. The solution will give the business users the ability to respond to market pressures as they are identified or before they happen. Through optimization of current structures coupled with responsive actions enabled by SAP, ConAgra will be able to exploit new windows of opportunity.

 ENABLEMENT BY HANA

Total Margin Management Powered by HANA

The SAP technology of this solution supports both real-time decisions and the execution of these decisions with a unified interface for multiple business areas, enabling insights within each area that all contribute to a more meaningful understanding of current and future conditions. HANA allows calculations to be extended to incorporate various business needs,

resulting in forecasts which build on the stock and work-in-progress conditions rather than on standard costs. This enables businesses to respond to forecasted challenges in real time as they are analyzed or predicted.

HANA's ability to manage huge amounts of operational data in memory allows for the implementation of scenario modeling without replication of unnecessary data, resulting in reduced costs for data storage and increased analysis capabilities. To prevent replicated data, TMM separates data into base data, which can be accessed by multiple scenarios, and user-specified adjustments for the base data, allowing for scenario customizations. Regardless of how many scenarios exist, the base data only needs to be stored once. Calculations for each user-customized scenario only have to reference a Virtual Data Model (VDM, see Section 2.2: Advanced Features) that is built from the combined base data and adjustments in order to produce results. Thanks to this single resulting baseline for all scenarios, the memory footprint required for each business model is significantly reduced and the scalability of the solution is maximized.

Due to the high speed of in-memory column scans in HANA, calculations which used to run as batch processes, reporting, and data analytics are possible faster than ever before. Thanks to the speed and storage features of HANA, the user is able to quickly perform iterative scenarios and what-if forecasting on large volumes of data. This increases the reliability of the margin forecast and produces a single version of the truth, while simultaneously creating a forecast that is risk-adjusted and applicable to multiple business areas.

9.3

Risk Analysis During Natural Disasters

Typically, risk analysis has a small window of opportunity during which decision makers need to make critical business decisions. Events that are difficult to predict, such as natural disasters or accidents, are typical examples of when decisions need to be taken within a short amount of time. For instance, the incident of the fake Associated Press tweet about an attack on the White House in 2013 made stocks quickly drop. Two minutes after the tweet was published, the Dow Jones Industrial Average index lost more than 100 points [CP13].

Providing short response times is an essential requirement for risk analysis applications in order to ensure that critical business decisions can be made in time. This requirement is yet more valuable in the context of natural disasters – e.g., hurricanes, earthquakes, volcano eruptions – as it can save people's lives. The problem of risk analysis for natural disasters is relevant for many organizations, such as hospitals, fire departments, transportation companies, and reinsurance companies. Real-time assessment of risks and implications of natural disasters allows such organizations to initiate proper measures at their necessary scale.

With the power of HANA, such analyses can be executed in seconds. In this section, we show how the HANA Spatial Engine can be used for hurricane damage prediction. SAP followed their co-innovation model with their partner Luciad, a company specialized in high-performance visualizations leveraging GPUs. We have jointly developed an application that enables almost real-time risk analysis for reinsurance companies. This application mainly shows two key strengths of the HANA Spatial Engine: its ability to process large amounts of data with sub-second latency and its capability to process and analyze business data and geospatial data within the same platform.

Providing short response times in the context of natural disasters is essential – as it can save people's lives.

Fast Predictions and Precise Computations for Reinsurance Companies

The problem of risk analysis is complex and involves, amongst others, the following challenges: (1) incorporating heterogeneous data sources to produce in-depth analyses, (2) performing complex calculations based on sophisticated mathematical and statistical models in real time, and (3) providing simple, yet powerful User Interfaces (UIs) and visualizations.

Risk analysis and damage prediction during natural disasters is vital for reinsurance companies. Potential damages influence the amount of claims reserves which reinsurance companies must hold, thus influencing their ability to invest remaining assets into other areas. Typically, risk analysis is complex and requires multiple hours of computation to complete.

💡 SOLUTION

Hurricane Damage Prediction

The Hurricane Damage Prediction application, powered by the HANA Spatial Engine, is mainly targeted at reinsurance companies. However, the application can also be used by insurance companies to optimize their policies, by telecommunication companies to manage service disruption, and local governments to assess infrastructure damage. For demonstration purposes, the data of Hurricane Sandy (Superstorm Sandy) was imported, which was the second-costliest hurricane in the history of the United States (US). This storm caused damage worth $68 billion in 2012 [Wal13].

After importing the data of the hurricane Sandy and overlaying it with public geospatial data for assets in the US northeast, Hurricane Damage Prediction can calculate its potential impact within one second. In order to estimate the impact more accurately, data from several heterogeneous sources is combined, including meteorological data, agricultural data, and data from the electricity grid.

Analysts at reinsurance companies are able to run an abundance of different scenarios to optimize their business. With the user interface shown in Figure 9.3, analysts can select the predicted storm data by selecting that storm from the list and using the time slider to simulate its effects over time.

Hurricane Damage Prediction provides detailed views of specific regions with which analysts can

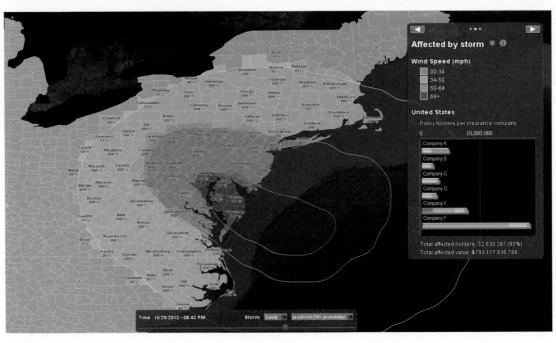

FIGURE 9.3
Simulation of the hurricane Sandy and its potential damage to the East Coast

estimate damage to assets located in that region. In the current co-innovation project, Hurricane Damage Prediction stores over 40 million assets with their geographic location and is able to calculate the expected damage within one second.

ENABLEMENT BY HANA

Real-Time Hurricane Risk Analysis and Prediction

Hurricane Damage Prediction leverages the benefits of in-memory columnar orientation for processing and storage of geospatial data. The HANA Spatial Engine can store all types of geometries such as points, line strings, and polygons. It can compute geospatial relations, e.g., intersections and containment, as well as distances between objects. The HANA Spatial Engine was designed to comply with major industry standards, in particular, ISO/IEC 13249 SQL/MM Spatial and OGC (Open Geospatial Consortium) standards. Relying on the existing body of standards enables interoperability with existing customer solutions and empowers customers to get the best solution from the combination of their Geographic Information Systems (GISs) data with their business systems.

Geospatial data sources are typically large in terms of data objects. Thus, spatial operations usually operate on large datasets and are computationally-intensive. For instance, sensors produce huge datasets in a very short time frame. HANA addresses this challenge with its columnar storage concept that brings computations and algorithms as close to the data as possible. The Hurricane Damage Prediction risk analysis utilizes spatial aggregation to display a heat map of assets. As there are tens of millions of assets, it is neither feasible to transfer the data to the client for every query nor to display it in a user-friendly way. Spatial aggregation in HANA can compute a reduced dataset on the fly and transfer valuable information to the user without overloading the network.

Another key strength of the HANA Spatial Engine is its ability to compute the intersections or unions of two datasets over a geospatial property. This is enabled by making heavy use of columnar storage mechanisms – such as dictionary compression, adjacent memory and optimized row orders – in addition to specialized algorithms that exploit these features. For instance, the HANA Spatial Engine uses various space-filling curve types to optimize the alignment of spatially close geometric objects within HANA memory. This smart layout of geometric objects speeds up memory scans by focusing on memory regions where adjacent objects are most likely to reside.

Smart organization of geometric objects in memory is another technique to speed up query execution, in addition to the fast full column scan. The combination of these techniques enable the HANA Spatial Engine to process huge geospatial datasets efficiently, which in turn enables real-time risk analysis over geospatial data.

CHAPTER TEN

QUALITY TIME AT WORK

E ngaging and empowering the user is the main objective of the User Experience (UX) of an enterprise application. It has to be immersive and satisfying. For this, the User Interface (UI) has to be responsive, intuitive, and provide optimized interaction patterns. Moreover, applications should enable users to retrieve the information they need to do their jobs, freeing them from dependencies on their co-workers. SAP achieves this by tailoring applications to specific target user groups.

Outline of this Chapter

10.1 Determining the Who, What, and Where of Consumer Behavior

Mobile network data presents a new opportunity for businesses seeking to better understand their consumer base and enhance their User Experience (UX). However, capturing, analyzing, and visualizing this data is challenging even for a typical medium-sized mobile network operator. From one such operator, we learned that 10 billion anonymized records are generated every day. Their challenge is to make this data easy to consume. As there are no suitable tools for processing this large amount of data, analysts can only get snapshots instead of a continuous flow of data in order to understand their mobile users.

Especially the marketing departments of companies have a large interest in such mobile data, as it is possible to reveal contextual data on human behavior and patterns without identifying individuals. To date, understanding consumer behavior has been relegated to panel tests and surveys reaching only a very small percentage of the population. This has frustrated many marketers because they have to make decisions on advertising strategies, location planning, and programmatic buying with data samples of less than 0.5% of consumers and outdated data.

By securely combining the data of mobile network operators with the possibilities of the HANA platform, SAP helps advertising departments understand consumer behavior on large scales. With SAP Consumer Insight 365, they can receive the anonymized mobile data of consumers in real time and study large populations with an unseen depth of insights. In order to process this large amount of data and easily provide answers for the questions of marketing end users, SAP created an intuitive and immediate UX that improves the quality time at work. Now, they have no difficulties answering marketing questions like how effective advertising was, what digital channels were most effective in bringing in foot traffic, and where the customers are coming from.

HANA can efficiently analyze billions of anonymized consumer data points from mobile networks.

Tapping into Mobile Operator Network Data

As a key part of their role, marketing analysts need to understand consumers:

> their socio-demographics
> places they visit
> areas they live in
> activities they engage in
> variation in activity preferences over time

This enables analysts to predict and understand user patterns and preferences so they can target their marketing accordingly. To better understand consumers, especially as society moves towards a mobile-centric lifestyle, analysts want to make use of behavioral data from mobile operator networks. By doing so, they can gain new marketing insights by accessing the wealth of largely unused, anonymized user data.

💡 SOLUTION

An Intuitive User Experience with SAP Consumer Insight 365

The SAP Design & Co-Innovation Center, together with SAP Mobile Services, developed SAP Consumer Insight 365. This is a cloud-based service powered by HANA that can easily aggregate and analyze the billions of anonymized consumer data points from mobile network operators according to local data protection regulations. With a focus on an intuitive UX, SAP Consumer Insight 365 securely enables insights based on mobile data and provides end users with market intelligence. This web-based application has won the Gold 2014 UX Award. The purpose of this section is to provide a feel for the intuitive User Experience that comes with SAP Consumer Insight 365 and show how businesses can gain insights to better understand consumer behavior from an end user perspective. As an example, the following figures (Figure 10.1 to Figure 10.5) present an analysis of foot traffic within a specific coffee shop. Foot traffic is critical for advertising attribution as it informs the business about how many people visit a place and how long they stay. After an advertisement campaign has been initiated, this can help determine whether the physical foot traffic has increased for any hour of the day, with an expected increase in select demographics. In addition, this analysis can also be used to determine programmatic buying and location planning.

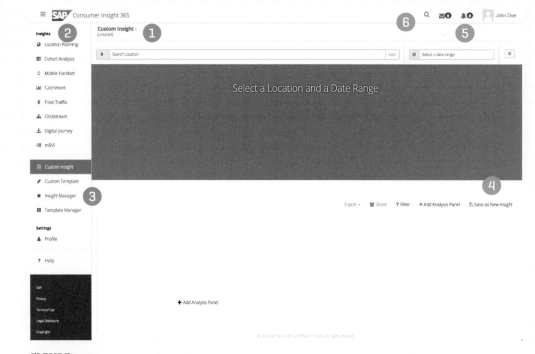

FIGURE 10.1
Starting screen of SAP Consumer Insight 365

The first screen of SAP Consumer Insight 365 (see Figure 10.1) is an empty Custom Insight screen ❶. Research with customers of SAP Consumer Insight 365 has shown users prefer the ability to create their own reports from scratch, as provided here. Other parts of the screen include:

> a menu of prepared insight reports ❷
> tools to allow building and management of reports ❸ that can be saved ❹
> social tools that can be used to share reports with others as well as receive notifications from the system ❺
> a search facility to find previously saved insight reports ❻

FIGURE 10.2
Entering location and date range to generate insights

Whichever report is being generated, the first step in Figure 10.2 is to enter a location ❶ and a date range ❷. Locations are input as a city name – in this case Guadalajara in Mexico. More than one location can be specified and compared with one another ❸. SAP Consumer Insight 365 uses Nokia Here® to help find locations through partial name searches as well as display the map itself.

FIGURE 10.3
Selecting the foot traffic insight for a specific coffee shop

Now, we look at foot traffic, which provides information about people visiting a place and how long they are staying. In Figure 10.3, the first step is to select Foot Traffic from the menu of prepared insight reports ❶. The next step is to search for and select a Point of Interest – in this case a coffee shop. The search found several coffee shops in Guadalajara. One of them needs to be selected ❷, and then the insight is generated ❸.

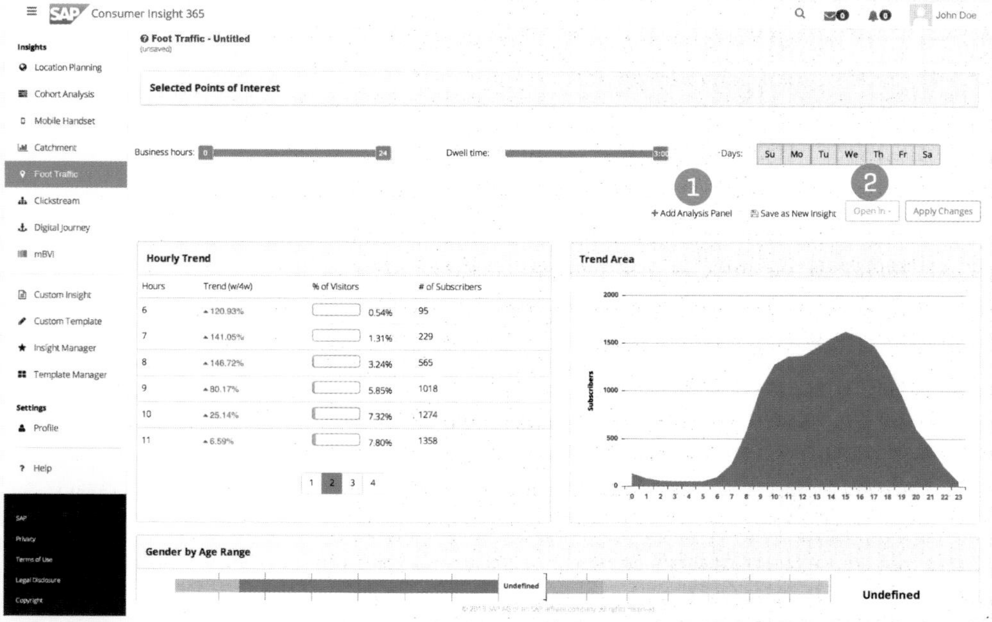

FIGURE 10.4
The generated insight panels for Hourly Trend and Trend Area

After the insight has been generated, Figure 10.4 shows the first two of the six insight panels. It provides data on hourly variation of the number of visitors in and around the coffee shop location. Note that:
> additional custom insight panels can be added ❶
> a new type of report for the same location can be opened using "Open In" ❷

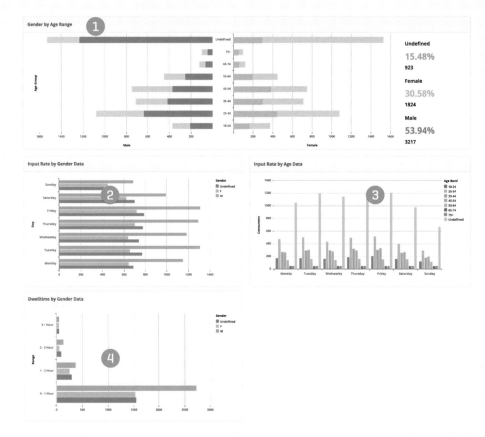

FIGURE 10.5
The generated insight panels: Gender by Age Range, Input Rate by Gender Data, Input Rate by Age Data, and Dwelltime by Gender Data

Figure 10.5 shows the remaining insight reports generated on the same screen (below the hourly trend graphics on the previous screenshot). The first report ❶ shows, for the period, the number of visitors by age range and gender. The age and gender of a large number of visitors is marked as unknown because these visitors are pre-pay mobile phone users who do not have to provide information on their age or gender to the mobile operator. The other graphics show how many visitors arrive each day of the week by gender ❷ and how many different visitors there are by weekday and age range ❸, and finally, the time visitors remain ❹. This kind of information can be valuable especially for retailers as they can use it in several ways:

> tracking how the number and type of visitors varies over time
> evaluating the effectiveness of an advertising campaign or coupon offer – did more people visit?
> determining if a location is a good place to open a new store
> deciding where to place an advertisement on a billboard

Yet perhaps one of the most valuable insights is the ability to research competitors by analyzing movement around their stores.

Reinvent the User Experience with SAPUI5 and SAP Fiori

The SAP Consumer Insight 365 User Interface (UI) and UX is designed by leveraging SAPUI5 and SAP Fiori. SAPUI5 is the SAP client UI technology that runs in any browser on any device (mobile, tablet, or desktop PC); SAP Fiori is the new UX for SAP software and applies modern design principles.

The SAPUI5 runtime is a client-side, web-based rendering library based on JavaScript, CSS3, and HTML5. With a rich set of standard, extension, and custom controls, it provides a lightweight programming model for desktop as well as mobile applications. It supports CSS3, allowing developers to adapt themes to their company's branding. Based on JavaScript and the open source jQuery library, SAPUI5 supports client-side features for Rich Internet Applications (RIA). SAPUI5 is also compatible with other JavaScript libraries.

SAP Fiori provides a personalized, responsive, and simple UX. It speaks a consistent design language and makes use of a common technical infrastructure. By blurring traditional computing boundaries and using interactive and attractive UI elements, Fiori provides a consistent end-to-end UX and can be used across all device types. Applications applying the SAP Fiori UX focus on the most critical and common activities and are designed around how people work.

10.2 All Key Performance Indicators at a Glance

The inventory of a typical large European retail company consists of several thousands of products that are grouped into interchangeable related categories. Large categories may contain thousands of products, and each product category is managed as an individual unit. With this categorization, category managers optimize the performance of categories as a whole (instead of single products). This enables companies avoid situations in which the promotion of one product cannibalizes the performance of another, and directs vendors to supply products which promote the growth of entire categories and maximize the retailer's profit. Category managers optimize their assigned product sections by planning inventory, setting sales prices, and negotiating purchasing prices from product vendors. For such negotiations, the managers need to determine the necessary amounts to buy and be able to identify and focus on the most profitable products. To do this, they collect information on specific vendors, check how profitable their products have been in the past, and determine their company share of the vendors' sales totals.

Key Performance Indicators (KPIs) help category managers maximize category performance despite the complexity of today's markets; with the power of HANA, all company employees such as category managers can generate and explore the KPIs necessary for their jobs. The Point-of-Sale Explorer is a prototypical solution built for category managers which enables optimization of category management and maximization of profits.

☐☐ POINT OF VIEW

Keeping an Overview of Thousands of Products

Nowadays, category managers have to rely on business analysts to retrieve necessary information and must request all relevant numbers in advance. It often takes analysts several days or weeks to extract the required data by modifying and running hand-coded SQL scripts on the enterprise database and compile the results in the form of slides or spreadsheets. Whenever additional detailed information is needed, category managers have to make a new request and again, wait for the response of the analysts. Category managers spend valuable work time planning and waiting for the retrieval of basic KPIs required to address a task, and there are only limited possibilities for them to refine the results which may be on already outdated information. In order to do their jobs effectively, what category managers need is an instant overview of KPIs such as the revenues and profits of their products.

💡 SOLUTION

The Point-of-Sale Explorer

At this point, the Point-of-Sale Explorer steps in. This intuitive and adaptable web-based cockpit presents a collection of the most important KPIs in a single view. For example, the Point-of-Sale Explorer can show revenues, profits, and market shares in a year-to-date or a rolling time period comparison.

The Point-of-Sale Explorer provides extensive filtering options. Category managers first choose a product category at the very top of the cockpit (see Figure 10.6). Then, they can refine the selection by retail chain and vendor. Their choice can be broken down even further to the individual product. In this case, the Point-of-Sale Explorer compares the KPIs

FIGURE 10.6
The Point-of-Sale Explorer provides category managers with a central view of KPIs for several thousands of products. The KPIs can be refined by store chain, vendor, and product, and can be displayed in tables, bar charts, or line charts.

for the selected product to the average KPIs of the product category. This comparison is displayed via tables, bar charts, and time series as appropriate. KPIs that differ strongly from their expected values are highlighted in red. The Point-of-Sale Explorer further provides additional information such as the companies' shares from the sales of a specific vendor and enriches them with external market share data.

In consequence, category managers do no longer have to start preparing for their negotiations weeks in advance. They do not depend anymore on other roles gathering information for them. Instead, the Point-of-Sale Explore delivers all necessary numbers intuitively, while also running on the latest available data. Category managers can slice and dice the data from which the KPIs are calculated, filter data relevant for their upcoming conversations, and access and explore all data right at the moment of preparing their negotiations. They can even do these tasks with mobile devices during negotiations.

This is what we consider a prime example of quality time at work: executing our core tasks without depending on support from others or being limited by outdated information.

The Point-of-Sale Explorer presents a collection of the most important KPIs for several thousands of products in a unified view.

User-driven Application Development with the HANA Platform

The key to the success of the Point-of-Sale Explorer is that it is specifically tailored to the needs of category managers. This way, it fulfills the category managers' comprehensive information needs while offering a simple, intuitive User Experience (UX). It grants the category managers independence from co-workers and empowers them to generate all information needed to get the job done on their own. This makes the Point-of-Sale Explorer a desirable solution.

The Point-of-Sale Explorer was put to the test using over 2.5 billion transaction data records

When developing the Point-of-Sale Explorer, SAP and the co-innovation partner once again experienced how operating on the finest granularity gives the highest flexibility.

from real business data. This data was enriched with over 30 million market research datasets and the master data of over 30,000 products. With this data, different scenarios for selecting the category, retail chains, vendors, and products were tested on the Point-of-Sale Explorer. Whenever the selection was changed, HANA consistently needed less than four seconds to generate all the various tables and charts depicted in Figure 10.6. Compared to weeks of requesting and combining these numbers without the Point-of-Sale Explorer, this is not merely a quantitative but also a huge qualitative step forward.

This excellent performance is the direct consequence of HANA's capabilities to process and analyze huge amounts of diverse data in real time. All steps – namely scanning tables, combining them, and aggregating the results – are extremely fast thanks to the in-memory columnar organization of the data. With a traditional database, a solution like the Point-of-Sale Explorer would simply not be feasible.

When a prototype becomes an actual product, it also must be viable, not only desirable and feasible. This means the overall development costs have to be balanced with the returns of the product. This is particularly important if the final product is not a general-purpose tool but targets a specific user group. It is for this aspect that HANA's greatest strength comes in – simplification. The complete Point-of-Sale Explorer co-innovation project, from design to implementation, was realized by six developers and two designers in three months. This was only possible because the Point-of-Sale Explorer makes use of HANA as a full-blown application platform.

The Point-of-Sale Explorer runs as a native HANA application. It uses the SAP Customer Activity Repository (CAR) framework. When developing the Point-of-Sale Explorer, SAP and the co-innovation partner once again experienced how operating on the finest granularity gives the highest flexibility. In traditional databases, materialized aggregates were still needed to achieve decent performance. Consequently, considerable amounts of development resources were spent on building and optimizing these predefined aggregates rather than improving the UX.

The category managers – our co-innovation partners – defined how they wanted to explore their product data. The co-innovation process leading to the Point-of-Sale Explorer was solely driven by the actual needs of the users. They chose the aspects for which the Point-of-Sale Explorer needs to be maximally flexible and those in which unnecessary complexity could be avoided. All development resources were put into realizing these requirements in the UI. This solution greatly leveraged HANA's ability to process any query despite its level of granularity. This is possible thanks to HANA's direct operations on the actual data and avoidance of materialized aggregates.

As the basic architecture and all functionality is provided by the HANA platform, resources are freed to focus on the needs of the individual user. This improves development efficiency and reduces the complexity of the infrastructure. With this, the software can target a specific user group and still remain affordable.

10.3
Navigation Through Enterprise Data

With S/4HANA, SAP presents a new concept of navigation through business data. With this approach, end users can navigate from business object to business object using their actual attribute values (referring to other business objects) as navigation paths. The columnar organization of the attributes allows a scan across various business objects independent of any application restrictions. This enables free navigation through any kind of data and improves workflows as people can find relevant information by themselves; external help for collecting work-relevant information is no longer required. HANA can provide a whole new User Experience (UX) that induces quality time at work – searching and browsing business data and taking action from insights is as simple as browsing the web.

◯◯ POINT OF VIEW

Context-based Navigation

The search for work-relevant information can be challenging. For example, employees in the procurement department are responsible for ordering materials. Therefore, they want to know which suppliers deliver a specific material, how a supplier performs, and whether a purchase order already exists for that material. In order to accomplish this task, users have to access the business application for the supplier master data, search for the transaction for the purchasing contracts, and access the respective spend analysis reports. However, current systems require a deep understanding of the system menu and knowledge of contexts assigned to each organizational area.

💡 SOLUTION

Browse Business Data

To enable simplified access to business content and allow users to use business objects beyond transactions and reports, HANA introduces a new navigation that shifts from transaction-centric to data-centric access. In order to find relevant business data, users start by entering free-text keywords into a search input field. With these, HANA scans across business content and finally presents a ranked result highlighting the hits, as expected on the web. HANA does not only consider structured information such as company codes or material identifiers, but also searches unstructured data such as attachments and text related to the business data. This makes it easy to find business content without accessing or knowing specific transactions. With this, users are able to browse business data the same way they browse the web.

Irrespective of differing application silos, the connected data can be located quickly. Once users find an interesting object, they can easily navigate to related ones or call the corresponding business functionality directly.

The navigation operates on top of a web of connected business entities called the Business Fabric, which has knowledge about all available business documents, their dependencies, and their relationships (see Figure 10.7). The Business Fabric is based on entities representing persons, organizations, or business objects. Each entity has, in turn, a set of attributes and can be visualized in fact sheets as well as searched, filtered, and clustered according to user needs. In order to create the Business Fabric, entities are interwoven with one another so that users can easily navigate along their edges. To ensure proper access rights, fact sheets as well as business entities automatically include all authorization checks.

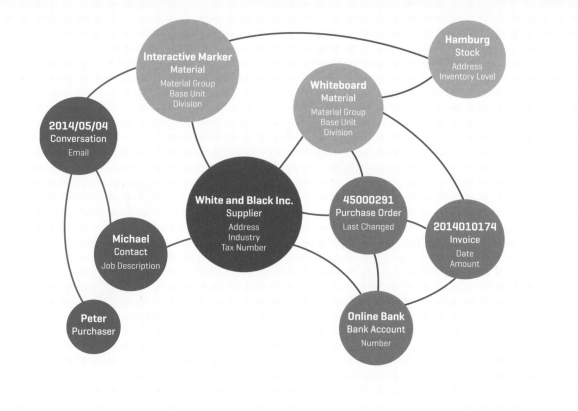

FIGURE 10.7
Users can navigate from one business object to another.

In the example depicted in Figure 10.7, the procurement department of a company wants to order more office supplies. Now, they can start by searching for suppliers that are able to deliver the desired material. After choosing White and Black Inc. as a possible supplier, a corresponding fact sheet with all important attributes and closely related entities opens. There is not only

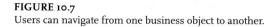

Users can now utilize business objects beyond transactions and reports.

general information such as main address, tax number, contact persons, and bank accounts, but also specific data about deliverable materials and already processed purchase orders. With this, users have all the necessary information of a single specific business entity at a glance. For more details, they can navigate to a connected business object, e.g., the ordered material. The corresponding fact sheet contains detailed information about the material and related information such as past invoices. If users are interested in even more business data, such as information about the inventory level of a specific stock, they can easily navigate to its related business object.

Simplification Enables the Business Fabric

With HANA, the new navigation feature can be implemented easily. The transactional and simplified data model serves as the basis for the Business Fabric. Entities are represented with views that are connected with associations (e.g., from a supplier to the issued purchase orders). Furthermore, they are enriched with key facts that are calculated on the fly, such as the annual spend of a supplier and the total goods receipts for a given purchase order. Finally, entities are connected to metadata to control search behavior and access rights. All this contributes to a comprehensive overview of an entity without the need to dig into analytical reports. To enable intuitive navigation, the business applications and transactions are linked to entities allowing users to launch and access them from each node of the Business Fabric.

In S/4HANA, the Business Fabric contains over 100 business objects (entities) which are linked and connected to one another. This translates to over 600 database tables that are covered by HANA views in order to realize this intuitive new navigation concept.

Thanks to the concepts of HANA, these new navigation concepts provide a new UX that enables quality time at work.

10.4
Driving Sourcing and Procurement Excellence Through Better Spend Visibility

Efficient spend management begins with enhanced spend visibility. Achieving efficiency across sourcing and procurement processes requires an accurate visibility of spending, suppliers, and related market trends in order to enable confident decision making. Yet, organizations find it increasingly difficult to gain a firm grip on their spending.

In this section, we present the journey from data-to-insights of one of the world's leading advertisement and communication services companies which leveraged the power of Ariba Spend Visibility on HANA. This new spend solution enables remarkably fast reporting, integrates different source systems, and automatically enriches spend data. Thus, users can achieve more in less time which, in turn, improves their quality time at work.

Spend Intelligence is the Key

A strategic holding company in the global advertising and marketing communication services space manages numerous leading national advertising agencies and a global network. This network consists of hundreds of marketing companies that service thousands of customers across over 100 countries. The scale of their operations and organizational footprint means that their spend data originates from a fragmented array of over 150 source systems and Accounts Payable (AP) applications. This spend data represents procurement transactions of over $90 billion, spread across a million suppliers. With this large amount of data tied up across so many different systems, there is often little or no insight for category managers concerning spending overlaps, supplier optimization opportunities, and savings leakages.

Gaining accurate spend intelligence is the key for increasing spend under management and improving negotiations with suppliers. Therefore, category managers need to answer questions such as, "What commodities are being purchased across the organization? Which suppliers are we purchasing them from? Are we paying the right prices? What commodities are our business units purchasing? Are there supplier linkages in our supplier base?"

SOLUTION

Answering Any Question with Ariba Spend Visibility

With an aim to consolidate and analyze its distributed spend data, the company deployed Ariba Spend Visibility – a cloud-based spend analysis solution powered by HANA. As an on-demand platform, Ariba Spend Visibility requires minimal implementation efforts and allows category managers to focus on the critical tasks of analyzing and planning, rather than on preparation activities. Equipped with the right information, category managers are able to expand their sphere of influence both at the individual company level as well as at the group level. The following examples show a range of reports and analyses which help managers answer important questions about spend data that are now feasible with Ariba Spend Visibility.

Purchase Price Variance By Commodity

| Pivot table | Chart | Dashboard |

▸ Applied Filters

[Data ▾]

Computer Equipment and Accessories ▾	UNSPSC (L4) ◈ ▾	Total Invoice Spend (USD)	Total Price Variance Cost (USD) ↓	Total Variance Cost Percentage (USD)	Total Invoice Count
Total		**694,882**	**86,147**	**12.40%**	**725**
Computers		**331,148**	**57,226**	**17.28%**	**81**
Desktop computers		62,712	29,324	46.76%	19
Notebook computers		180,273	16,293	9.04%	25
Computer servers		27,601	10,669	38.65%	7
Personal computers		60,562	939	1.55%	30
Computer data input devices		**339,168**	**23,461**	**6.92%**	**602**
Scanners		37,963	22,221	58.53%	6
Computer mouse or trackballs		1,342	724	53.94%	23
Bar code reader equipment		283,994	275	0.10%	36
Keyboards		15,870	241	1.52%	537
Computer printers		**17,656**	**5,355**	**30.33%**	**36**
Laser printers		17,656	5,355	30.33%	36
Computer accessories		**6,910**	**105**	**1.52%**	**6**
Port replicators		1,270	105	8.27%	5
Docking stations		5,640	0	0.00%	1

FIGURE 10.8

Example report for purchase price variance by commodity. Computers and printers especially show notable differences in their purchase prices, with a potential for savings.

Figure 10.8 shows commodity price variances. This report displays invoice spend, price variance cost, and count for top-level commodity categories in an aggregate view. With this, users are able to analyze differences in commodity prices by supplier over time to identify saving opportunities (here the supplier is hidden in the applied filter option on the top). For example, desktop computers have a price difference of nearly 50% for this specific supplier and the reason needs to be clarified by category managers. Therefore, they can dig into more details and look at single line items in another report.

Purchase price alignment cost is the lost savings of different business units purchasing the same items, from the same supplier, at varying prices. This helps a category manager analyze if the company is losing savings on each commodity by not aligning purchase prices with individual suppliers. Going deeper, category managers can analyze off-contract spend by commodity to determine if the group companies are failing to realize savings through maverick spend, and thus, identify opportunities to further negotiate contracts in specific commodities.

Using Opportunity Analysis reporting, users can determine categories that need order consolidation. They can also find commodities which accounted for a relatively large percentage of invoices with varying prices across multiple suppliers.

By analyzing supplier ownership hierarchies and overall expenditure per supplier, a category manager can determine whether users are purchasing the same commodities from suppliers that are identified as separate units in the source systems, and therefore, simplify analytics due to compression of the reports.

Category managers can also create a critical supplier profile and financial analysis to assist in risk assessments and risk management by using supplier financial ratings and examining supplier credit and revenue ratings.

These reports are just a small selection of what is now possible with Ariba Spend Visibility. This gained flexibility of analyzing spend data in combination with the fast reporting of HANA is the most important impact on the strategic holding company. Category managers are able to identify quick-hit opportunities which could yield savings within existing contracts, build a stronger sourcing pipeline by developing opportunity lists, and improve compliance adherence by monitoring all global spend activities.

Furthermore, leveraging HANA's benefits, the overall reporting times decreased by 80%. This in turn led to an increase of 74% in the unique number of reports run and an increase of 134% in the total number of reports run compared to the pre-HANA reporting days. The much shorter response times have attracted many users to exploring the analytical options. We completely had underestimated how performance shapes the behavior of users. The increased usage is an impressive testimony to how HANA can improve the relevance of applications.

> *Using HANA, the overall reporting times decreased by 80% and the total number of reports increased by 134%.*

Enriching Spend Data within the Cloud

Using the power of in-memory computing, the platform offers category managers the possibility to run custom spend analytic functions and create reports spanning thousands of commodities and suppliers. Users now have the flexibility to run ad-hoc reporting tasks based on custom parameters, as well as leverage out-of-the-box prepackaged reports.

Ariba Spend Visibility does not only access existing spend data within a few seconds, but rather enriches and combines it with additional sources of information. In this way, users can ask more difficult questions and reveal unknown relationships between spend and supplier data that were otherwise difficult to understand and visualize. Running in the Cloud, the completely automatic enrichment process looks as follows:

1. Classification
As the first step, the extracted data from multiple Accounts Payable (AP) systems is uniformly classified using a combination of industry-specific and custom taxonomies. This step ensures that all commodities are laid out in a hierarchical format, allowing for drill-downs.

2. Business Insights
Subsequently, all supplier records are enriched using the unified Dun & Bradstreet (D&B)

Business Insight database which includes data for over 230 million businesses worldwide. The supplier data is enriched with industry codes, parentage information, business financials, and other supplier related information. market information

3. Market Information

The enriched and classified data is then blended with related market information such as price index information which enables category managers to measure price movements and volatility. Ariba Spend Visibility also offers insights on probable sourcing savings for multiple commodities using historical sourcing data.

4. Reporting

Finally, users can slice and dice their enriched spend data based on a number of prebuilt and user-defined parameters. HANA enables users to run complex and multi-parameter-based reports without any limitations to data size. Thus, the solution can be easily adapted to the varying reporting requirements of end users.

CHAPTER ELEVEN

SOLVING THE UNSOLVABLE

I n the past decade, the evolution of hardware – in particular the steep scale-up in main memory and the massive parallelization within single systems – has enabled us to do things that were simply impossible before. We can now process huge amounts of data in almost zero time. Big Data comes in many forms, be it structured enterprise data (e.g., master data and transaction data), sensor data (e.g., geospatial data and sensors in the Internet of Things), or unstructured data (e.g., websites and social media). The capability to process and combine all three of these data types is one of HANA's unique features.

Outline of this Chapter

SECTION 11.1 **SMART VENDING** Vending machines present a yet untapped valuable source of information for their operators. Smart vending machines, as part of the Internet of Things (IoT), represent the future of connected consumer goods management. With HANA, vending machine operators are able to remotely manage inventory and maintenance while displaying targeted advertisements and special offers to their customers.

SECTION 11.2 **REMOTE SERVICE AND PREDICTIVE MAINTENANCE** In the IoT, consumer products are equipped with sensors and send their status to the vendor via the Cloud. With HANA's predictive capacity and the SAP Cloud, we can see how predictive maintenance for every machine is possible today.

SECTION 11.3 **EMPOWERING LOGISTICS HUBS** Over 900 organizations operate at the port of Hamburg. Every year, this logistics hub manages 10,000 ships and more than nine million containers. The smartPORT logistics project provides a comprehensive IT platform that allows the logistics hub to increase its capacity without expanding its physical facilities.

SECTION 11.4 **TRACING GOODS AROUND THE GLOBE** Medicinal counterfeits are a severe medical and financial problem for the European pharmaceutical supply chain. With HANA, it now becomes possible to process the immense amount of data necessary to identify, trace, and authenticate prescription medicinal products.

SECTION 11.5 **UNDERSTANDING THE OPINION OF THE DIGITAL SOCIETY** The image of a company is increasingly shaped by social media. Traditional news sources are complemented or even replaced by Internet mediums such as blogs. The BlogIntelligence research prototype supports mining, analyzing, modeling, and presenting the immense data collections of blogs in order to assist companies in understanding their customer base.

11.1

Smart Vending

One of the main challenges companies face when selling Consumer Packaged Goods (CPGs) via vending machines is the limited direct interaction with customers. Without the ability to communicate with and understand the behavior of their customers, it is difficult for operating companies to manage inventory and production. Given the information delay (or even lack of information) inherent to the current company-to-retailer-to-consumer model and the issues that arise thereby, it is the goal of producers to find a way around such shortcomings.

The vending machine is a direct-to-consumer method of providing a product. However, even today in the era of ubiquitous Internet usage, the machines are as mechanical and old fashioned as they were when originally designed. They do not report if they are short on a stock item, if they are selling well (or poorly), or if there has been a malfunction. Yet, all of these things have a direct impact on the bottom line of the operating company.

POINT OF VIEW

Understanding Customer and Machine Needs

Operating companies of vending machines would like to be able to utilize the wealth of information collected and generated by their machines. Such information about consumer buying habits and machine status can provide insights which allow companies to optimize their offerings and machine maintenance to maximize profit. Operating companies encounter several challenges for achieving this goal: current vending machines are primarily cash-based, meaning that their users are often anonymous; maintenance and restocking is still done via scheduled on-premise checks; and finally, static door fronts and small displays limit possible advertising and interactivity. In order to meet these challenges, operating companies want to not only efficiently gather and store data from customer interactions, but also understand and utilize this data so that they can draw valuable business insights. Transforming such data into a usable format requires an understanding of Big Data analytics and high-volume processing to deal with the large amounts of data produced. Being able to collect, store, analyze, and understand this data would allow operating companies to reduce inefficiencies in the stocking and maintenance process, lower warehouse and transportation costs, connect with outside channels to publicize promotions, and even suggest personalized discounts and bundles to customers through an interactive display.

💡 SOLUTION

Consumer Packaged Goods in the Internet of Things

The SAP Connected Trade Assets team worked with potential initial customers to design and build a ready-to-use solution that addresses the problems faced by CPG companies such as operators of vending machines, and similar corporations in the modern connected landscape. As consumers increasingly interact with the world around them via smartphones (using them to find products, give direct or indirect feedback, and often, purchase products), smartphones present the ability to be used as a means of payment via web services, Near Field Communication (NFC) wallets, and other competing technologies, while also enabling personalized interactions.

To solve the challenges faced by vending machine operating companies, the SAP Connected Trade Assets team combined a smartphone wallet application with a smart vending machine, utilizing the data collection and analytical capabilities of HANA (see figure 11.1). While in the past, customers had been entirely anonymous to vending machines, now they can interact with the smart machines via a smartphone application that can receive personalized offers and notifications. For the operating companies, the new smart vending machines further simplify business by automatically communicating inventory levels and maintenance requests without human intervention. SAP Connected Trade Assets enables operating companies to optimize their businesses thanks to increased ability to understand their customers and maintain their machines by addressing the aforementioned challenges in the following ways:

› consumer information is captured through interaction with the machine and the smartphone application. This information is collated and analyzed in real time in HANA, and builds a picture of the consumers, their likes, dislikes, and trends. Finally, based on the collected information, the machine can motivate customers to purchase more by communicating product discounts, marketing campaigns, and even individualized messages.

› the smart vending machine transmits its inventory levels in real time. It can indicate a low stock number on demand, and as a result, a distributor only needs to visit machines that are low on supplies. The automatic, real-time transmission of machine inventory reduces costs for warehousing, transportation, and personnel.

› the large touch screen of the smart vending machine can be remotely configured to promote specific products or display special offers. The increased interactivity of the smart machines promotes more meaningful customer interactions as well as providing the ability to connect with outside promotional offers and channels.

› cash is no longer an issue as the smart vending machine allows the user to pay via web services, Near Field Communication wallets, or the smartphone app. This expands the possible customer base by simplifying and expanding payment methods.

FIGURE 11.1
Smart vending machines and their touch-enabled User Interfaces (UIs)

›the smart vending machine transmits its operational status and sensor data to the distributor. As a consequence, the distributor is immediately informed when malfunctions are detected and can schedule maintenance.

🔑 ENABLEMENT BY HANA

The Transition into a Whole New Product Channel

SAP Connected Trade Assets relies on the Smart Data Streaming feature of HANA to collect telemetry and sales data from smart vending machines. Smart Data Streaming provides complex real-time event processing within the smart vending landscape by performing three basic functions within HANA: it receives and consolidates the incoming stream of events from smart vending machines; it normalizes the incoming events and persists some or all of them in the HANA in-memory columnar storage; and finally, it monitors and analyzes the incoming data in real time to watch for conditions or patterns that warrant an immediate response. Data is subsequently stored in the HANA Cloud, which provides real-time analytical access to the data at the highest level of granularity. With the support of HANA, CPG companies can start their transition into a whole new product channel.

While the SAP Connected Trade Assets solution exemplified here primarily focuses on the CPG market, it has already been expanded to cover the wholesale, telecommunications, over-the-counter pharmaceuticals, retail, and digital signage industries.

11.2

Remote Service and Predictive Maintenance

Fifteen years ago, the first machines were connected to maintenance applications. Companies connected their products in order to collect sensor data to optimize service processes and ensure higher availability. Engineers could manually perform remote machine diagnostics by observing and evaluating the collected sensor data. This led to higher service quality through faster response times, improved root cause analysis, and lowered service costs due to fewer technician visits.

However, the requirements for the remote diagnosis of machines change as the quantity increases from several hundred to hundreds of thousands. In 2015, the Internet of Things (IoT)

HANA is able to efficiently deal with large amounts of sensor data.

is estimated to already consist of over 25 billion connected devices with a rising trend of 50 billion by 2020 [Eva11]. As the former manual diagnostic performed by engineers does not scale with these numbers, this technological shift requires more automation. For this reason, fast services are required to handle the large amounts of sensor data received, notify engineers about all possible deviations from normal operation, and predict on-site maintenance demands. In addition to this, it is no longer practical to have a separation between the maintenance application for remote operational diagnostics and the business view for service management.

In this section we present how HANA is able to efficiently deal with large amounts of sensor data to enable low-cost remote service and predictive maintenance. The ability of HANA to process structured and unstructured Big Data, together with its predictive analytics functions, allows companies to work with the increased data produced from the higher amount of sensors per product. In order to provide further innovative maintenance concepts, SAP combines these abilities with our Customer Relationship Management (CRM) system, and by doing so, provides seamless integration into existing service processes. With this combination, there is a high potential to save costs by reducing unproductive downtimes of expensive production and transportation equipment, as well as lowering repair and overhaul costs.

Permanent Monitoring and Diagnosis of Remote Machines

SAP works with a co-innovation partner who manufactures fast-turning machines. For the past ten years, the company has offered an exclusive diagnostics service for its premium line of products. Over 100 machines located at their customer sites are connected via the Internet to a standalone remote diagnostic application. Each machine sends up to 120 sensor values each second. Engineers manually examine the sensor data of a machine and its generated alarms, and write a quarterly report based on their experience. These operation experts in remote machine diagnostics were entirely occupied with analyzing the data and writing reports in order to provide the premium service.

This service was highly appreciated by the customers, and so, the company saw an opportunity to increase customer loyalty and service revenues by offering the remote diagnostics service for all of their products. As each machine has an average life span of ten years, this would lead to a much larger pool of machines to monitor remotely. For this reason, it is clear to the company that the current process of manual remote diagnostics and report writing does not scale for this new business model.

Connecting Machines to the Cloud

In order to solve this challenge, the company decided on a cloud-based solution with HANA. This new remote service and predictive maintenance application fulfills the following expectations:

› the company sees a cloud-based solution as standard for the industry, and expects lower IT costs compared to an on-premise solution run by its own IT department.

› the solution fully integrates and combines sensor connectivity, device management, remote diagnostics, reporting, and service processes managed by the SAP CRM system.

› the remote machines connect with the application and send sensor data in high volumes. In the application, engineers are able to define rules for issue detection using the incoming sensor data and execute these rules in real time. For example, the solution automatically monitors critical vibrations of machines and reports outliers as soon as possible.

> the application automatically generates the required reports in order to limit manual work by engineers.

> to enable predictive maintenance for machines, the application requires access to all machine data from past years. With this data, it can run cross-machine analytics to find new patterns and create rules for detecting new deviations from the norm. This algorithm is able to reveal knowledge such as a specific motor requiring additional lube oil every six months in order to decrease machine failure situations.

Based on these requirements, SAP built a native HANA application using the HANA Cloud Platform (HCP) with IoT Enablement for connecting the machines to HANA. Each machine is shipped with an industrial PC for process control and data acquisition that runs an SAP IoT Connector to receive machine signals and send the data into the HANA Cloud. To store the sensor data in the Cloud, the HANA Big Data Platform is used. Sensor data and events are processed and stored in memory for a short retention period, typically days or weeks. When data arrives in HANA, rules are executed in real time to either filter out minor outliers or create new alarms. In addition, SAP IQ – a disk-based, petabyte-scale, column-oriented, relational database system – is used for long-term data

storage with a retention period of months or years. The stored data is still available in HANA through Smart Data Access (SDA). This solution is cost-effective for long-term storage of historical sensor data.

With this application, remote service engineers are automatically notified about machines that require diagnostics. Engineers employ a specialized visualization for this purpose, as shown in Figure 11.2. In this example, the vibration analysis of a remote turning machine (yellow line) reveals that at high frequencies, the amplitude exceeds expected thresholds (green line and red dots). Thanks to the speed of HANA, even large datasets are visualized quickly, and with the help of other capabilities such as predictive algorithms, engineers can also address domain-specific problems such as the analysis of exceptional vibrations. In this manner, engineers can identify which components of a machine might have a problem, zoom into the data, and identify the root cause of an alarm.

When the data basis is sufficiently comprehensive, predictive algorithms are applied to the sensor data in order to provide a predictive maintenance scenario. Such scenarios can identify patterns that lead to machine failures in the historical sensor data and immediately recognize such patterns in the incoming sensor data. With the additional CRM integration, this application can further create work orders and maintenance appointments before a machine's possible downtime. This leads to better planning of maintenance resources and prevents expensive repairs.

FIGURE 11.2

A remote turning machine shows at which frequencies it violates amplitude thresholds. The frequency band with violations (red dots) enables insights concerning the vibrating machine component.

The Platform That Enables Remote Services

Remote service and predictive maintenance is a complex application built upon a set of requirements that must be addressed together. The HCP provides the means necessary to build powerful cloud solutions. This cloud application is built with the help of Fiori, and the resulting web application can be used on any browser regardless of whether it is on a desktop or a tablet.

HANA's columnar storage makes it possible to design generic data models that are flexible to many machines and allow for dynamic adaptation during the lifetime of a machine, e.g., when a sensor is added. Using a Virtual Data Model (VDM), the data can be immediately transformed into different representations for specific analyses. Furthermore, VDMs are able to integrate business data and create new insights, e.g., costs of maintenance with respect to specific machine conditions. In this way, machine parts or configurations that lead to higher maintenance costs can be identified.

For defining and executing rules on incoming sensor data, the HANA Rules Framework is used. Rules combine various sensor values in order to create notifications concerning specific machine conditions (e.g., raise a warning if temperature, pressure, or motor amperage exceed specified thresholds). Rules are defined by end users and uploaded to the HANA Rules Framework. These rules are then executed in real time on the incoming sensor data and can reduce unnecessary downtimes by providing early detection.

The HCP offers further services that are required for the remote service and predictive maintenance application. Using the HANA Cloud Connector, the existing on-premise systems such as CRM are connected to the cloud application. The SAP

HANA's columnar storage makes it possible to design generic data models that allow for dynamic adaptation during the lifetime of a machine.

Cloud Identity Service provides user management and single-sign-on capabilities. Furthermore, customers and partners can easily extend the solution by adding additional HCP applications that access the same data store.

11.3

Empowering Logistics Hubs

Contemporary supply chains are becoming increasingly global and interconnected. Between 1995 and 2007, the number of international companies has more than doubled from 38,000 to 79,000 [IBM10]. Supply chains must accommodate both the increasing number of participants as well as constantly expanding product portfolios. This puts supply chains under pressure, and requires a high degree of flexibility to allow them to cope with the pace of the market.

Critical nodes within supply chains are logistics hubs because they play a central role in improving the flow of traffic and goods. For example, the operators at the port of Hamburg manage 10,000 ships and greater than 9 million containers per year as well as 40,000 trucks every day. Their operations are fundamental for reductions of logistics costs and increases in the speed of supply chains.

One of the main issues faced by logistics hubs is the need to cope with the increased flow of goods. For example, only 30% of the time an average truck spends in the harbor area is driving. The remaining 70% is spent waiting at terminals, bridges, and in traffic jams. The traditional way of solving this issue was through the expansion of logistics hub facilities. However, as these are mostly located within city centers, both geographical limitations and property costs make this approach difficult.

The research project smartPORT logistics between Hamburg Port Authority (HPA), Deutsche Telekom, and SAP set out to address this issue [Ham14]. The mission of the project was to help all users of the HPA logistics hub, i.e., over 900 organizations and their personnel, improve the efficiency of their processes. For the HPA, where 145 million tons of goods were moved in 2014, there is a substantial economic value to be derived from even small efficiency gains when it comes to logistics and traffic. More efficient processes allow for further growth of logistics hubs within the same geographical limits.

POINT OF VIEW

Fostering a Transition to Digital Services

The first step towards sustainable growth for logistics hubs is IT-powered innovation. Figure 11.3 presents the trade-offs faced by logistics hub professionals. Based on these trade-offs, we can derive the following critical requirements necessary to improve the efficiency of logistics hub operations:

> the lowest inventory levels cannot be achieved without transparency for locations and destinations of moving assets (trucks, containers, etc)
> minimizing the demand uncertainty requires extensive interaction between independent business partners within a single hub and across different hubs
> keeping up with an on-time delivery implies a better, bilateral control of goods, transport media, and emergency personnel

These requirements can only be addressed through the real-time optimization of logistics hub operations.

SOLUTION

smartPORT logistics

All partner companies initiated a project to optimize traffic and logistics operations that allow larger quantities of goods to be transshipped within the HPA controlled port. Road capacity within the Port of Hamburg is restricted and options for modifying the roads to accommodate additional vehicles are limited. Therefore, the port urgently required an efficient traffic management system to continue growing.

The smartPORT logistics pilot project resulted in a comprehensive IT platform that incorporates mobile applications and makes it possible for traffic information and port-related services to be accessed from mobile devices. While T-Systems provides the technology for vehicle-related, real-time services and an application for smart devices, SAP contributes the SAP Connected Logistics solution running on the SAP HANA Cloud. The complete system supports mobile personnel, such as truck drivers, with real-time information from HPA's Port Road Management system. The information is delivered in the form of up-to-date, personalized messages covering the area in and around the harbor. The provided information, originating from varying sources, is aggregated in a private cloud and includes specific port and road traffic information, port infrastructure (e.g., blocked bridges or terminals) information, as well as information about the availability of parking spaces. Using the system, the participating freight-forwarding companies are able to track their transshipping orders in real time.

Allowing the freight planners and truck drivers to communicate more effectively meant that there was a significant reduction in the time required to respond to traffic disruptions. For example, typical incidents included bridge closures, which required redirection to alternative truck-holding

Sustainable Growth

Highest Customer Service Levels	**VS.**	Lowest Inventory Levels
Minimize Demand Uncertainty	**VS.**	Maximize Customer Satisfaction
Minimize Transportation Cost	**VS.**	Maximize On-Time Delivery

FIGURE 11.3
Constant challenges for supply chain professionals in logistics hubs

SAP Connected Logistics, based on the SAP HANA Cloud, integrates all relevant information from multiple data sources.

areas. The smartPORT logistics project resulted in a decrease in the truck waiting times and fewer traffic jams within the port area and on approaching roads. This led to an increase in productivity greater than 12% only through the use of smartPORT logistics [Mei14]. Moreover, through the extension of the smartPORT logistics pilot project to include trains, bridge intelligence, and multiple competency areas, it can be expected that the total throughput can be increased by up to 50%.

Insights and lessons learned from the smartPORT logistics project were directly transferred into the SAP Connected Logistics product.

⚷ ENABLEMENT BY HANA

Optimized Hub Operations with SAP Connected Logistics

HPA and other hub operators are keen to monitor key figures and characteristics about traffic and infrastructure in order to improve the efficiency of port operations. Collaboration among business partners is a key success factor (see Figure 11.4) – with the SAP HANA Cloud forming the underlying platform.

Within the pilot phase, various analytical services have been defined in accordance with the demands of HPA. Through geofencing and telematic service provided by T-Systems, port authorities can monitor how many vehicles are located in predefined areas at a time, as well as track the maximum number

Real-Time Road Traffic Control System

Smart Operations Platform

Port Road Management and Operations Back-end Systems

SAP Connected Logistics operated by SAP HANA Cloud Platform

Freight Forwarders and Truck Drivers

Parking Space Providers

Business Network of 950+ Connected Business

FIGURE 11.4
Project overview – today's participants and future connected businesses

of trucks in select areas – all in real time. Shipping companies and freight forwarders can access the same service, restricted to their fleet.

Another important service is dedicated to the problem of parking spaces, which at certain times become greatly overloaded. In order to solve this problem, real-time tracking of parking space availability and automatic dispatching were implemented. Finally, a real-time service for incident tracking, e.g., traffic jams and blocked bridges, was developed.

SAP Connected Logistics, based on the SAP HANA Cloud, integrates all relevant information from multiple data sources including location and partner-dependent information. The total amount of information processed by SAP Connected Logistics poses both challenges and opportunities. The major challenge comes from the fact that the total number of objects to be tracked increases if we include ships, containers, bridges, and trains across multiple logistics hubs. HANA addresses this challenge in two ways. First, thanks to its columnar layout and data compression, it is able to store and manage all necessary data in memory, providing real-time answers. Second, through the Smart Data Streaming component, HANA can answer queries while the information is still on the move, i.e., before it is permanently stored. Collecting all data in HANA allows users to obtain a holistic view of the situation within a single hub and across hubs. With all this information at hand, logistic hubs can further optimize their operations on all levels of granularity, resulting in increased potential for higher throughput.

11.4
Tracing Goods Around the Globe

As global trade increases, opportunities for producing and selling counterfeit products also arise. The Organization for Economic Co-operation and Development (OECD) estimates that the trade volume of pirated and counterfeit goods (excluding domestically produced and consumed products and pirated digital products) reached $250 billion in 2009 – an equivalent of about 2% of the world trade volume [OEC09]. This development has also led to an increased number of counterfeit medicinal products in the European pharmaceutical supply chain. As a result, the pharmaceutical industry launched the operation MEDI-FAKE, involving customs authorities in all EU member states. The operation resulted in more than 34 million fake drug tablets being detected at the borders of the European Union between October and December 2008. Detected counterfeits would have put lives in danger as they included vital medicines such as cancer medications, antibiotics, and painkillers which contained false ingredients, incorrect doses, or other harmful substances.

The European Commission has called for new measures to combat the medicinal counterfeit problem that ensure the identification, traceability, and authenticity of prescription medicinal products. To provide a sustainable prevention mechanism against medicinal counterfeits, a system for tracing medicinal products through their lifecycle must be implemented. The HPI, in cooperation with the SAP Innovation Center Potsdam, has developed a research project built with HANA to establish such a prevention mechanism.

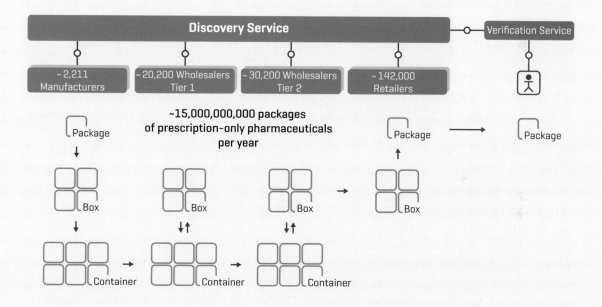

FIGURE 11.5
Overview of the European Pharmaceutical supply chain

Billions of Medicinal Packages to Trace

To ensure medicinal product validity and keep counterfeits out of the market, Unique Product Identification (UPI) must be provided for each prescription drug and its packaging. These UPIs help form hierarchical UPI relationships amongst transported pharmaceuticals, identifying the movement of prescription medicines. Readers must also be set at strategic points in the supply chain enforcing documentation of UPIs and additional lifecycle metadata. Moreover, traceability and authenticity verification should be based on the information provided by the readers across the entire supply chain.

 SOLUTION

Track and Trace Service for Medicines

The pharmaceutical supply chain has multiple steps which complicate the tracking process; to ship medicinal products, manufacturers aggregate packages of pharmaceuticals to a box and multiple boxes to a container; wholesalers then disaggregate containers and reaggregate the containers again so that they can be forwarded; and finally, these packages are sent to the destination retailers and customers (see Figure 11.5). In this supply chain, readers must be positioned to securely record each step of the pharmaceuticals in their transportation and packing lifecycles. This information contains sensitive data and cannot be openly shared, yet can be used to ensure the validity of medicinal products and identify counterfeit medicines.

In order to address the challenge of the medicinal counterfeit problem, a special information system – the Hierarchical, Packaging-aware Discovery Service (HPDS) [Mül13] – must be provided so as to identify all relevant information for a specific product of interest. The HPDS is a superordinate entity that supports and coordinates inter-organizational collaboration and information retrieval, and allows for a new communication pattern between the IT systems of requestors and supply chain partners such that a minimal number of messages has to be exchanged. The HPDS also enables the use of thin clients, such as Point of Sale (PoS) terminals and mobile devices, to easily interact with the service and to simultaneously ensure that companies retain full ownership of their data. The basic functioning principle behind the HPDS is as follows:

1. retrieve a query from a client (e.g. a product scan in a pharmacy)
2. identify all relevant supply chain partners

3. query all identified supply chain partners' IT systems in parallel
4. aggregate their responses
5. respond synchronously to the client request

The resulting complexity of the discovery service is handled by two algorithms. The first algorithm is an efficient, heuristic product search algorithm, and the second is a filter algorithm which processes product data returned by the search algorithm and evaluates whether the information is relevant for a given product. In addition to the communication protocol, the search algorithm, the filter algorithm, and the data management are optimized for column-oriented, in-memory databases with dictionary encoding. This opens the opportunity to handle the data volume of 15.6 billion packages of prescription pharmaceuticals entering the European Union market per year, which translates to about 35 billion events in the supply chain that have to be processed by the discovery service.

HANA opens the opportunity to handle the data volume of 15.6 billion packages of prescription pharmaceuticals entering the European Union market per year.

⌐ ENABLEMENT BY HANA

Hierarchical Packaging-Aware Discovery Service

The HPDS was implemented using HANA as the platform of choice. HANA provides the necessary capability to store and process information concerning the 30 billion packages of pharmaceuticals produced every year within the European Union, and its scalability allows the handling of the real-time workload originating from all European pharmaceutical supply chain participants. The evaluation showed that in order to run an HPDS for the complete European pharmaceutical supply chain, only 1.53 terabytes of main memory are necessary. This is possible thanks to the columnar data representation and the dictionary-based compression of HANA which compresses the original data (9.66 terabytes) by a factor of 6.3. The speed of the HPDS system is determined by the computational complexity of the search algorithms. By exploiting parallelization, HANA can reduce the search time proportionally to the number of available cores in the underlying hardware. This allows for an increase in the throughput of the whole system and reduces the hardware requirements.

In the future, the involved partners plan to expand to further relevant areas such as effective support of recalls, company-spanning supply chain optimizations, and pattern recognition in supply chains. As the presented discovery service explicitly integrates changes in packaging hierarchies, this approach can easily be mapped to the bill of material problems, e.g., to identify all parts of an airplane and their history at an arbitrary point in time.

BlogIntelligence is able to process, analyze, and present a snapshot of blog data from the World Wide Web in nearly real time.

11.5

Understanding the Opinion of the Digital Society

With the widespread adoption of social media, its inherent value also gets more and more important for business. As millions of social media users often discuss companies, products, and services, it has become essential for businesses to understand the opinions of the digital society. Blogs are one of the most used applications for expressing opinions in the World Wide Web. Publicly available blogs can be easily accessed, written, and read by anybody. With an immense circulation of over 200 million blogs around the world, they create a representative sample of public opinion, especially in the young customer segment.

Even if blogs present a valuable source of information, their comprehensive observation is challenging. To get an overall picture of customer opinions, companies have to explore the huge number of blogs and discussions distributed all over the web. Traditional tools are stretched to their limits with respect to mining, analyzing, and combining this immense amount of structured and unstructured information in an efficient way.

For this reason, new ways are required that allow companies to gain deeper insights into the opinions of the digital society. In this section, we present BlogIntelligence as a research project that is being developed by the HPI in cooperation with the SAP Innovation Center Potsdam. Based on HANA, this tool is able to process, analyze, and present a snapshot of blog data from the World Wide Web in nearly real time [MBBH15]. As a result, companies can easily ask questions about their products and get answers from publicly accessible feedback. Nowadays, this information is an indispensable tool for companies in order to react to shortcomings and control business.

∞ POINT OF VIEW

Exploring Blogs in the World Wide Web

Public opinion and company image play a major role especially in marketing, sales, and product design. Social media, and blogs in particular, can enable companies to detect and react to customer issues before they damage the reputation or even cause a loss in profit. Unfortunately, due to the growing number of blogs and the enormous amounts of included information, manual exploration and understanding is nearly impossible for humans. Therefore, intelligent social media analytics tools have to enable users to understand, for example, what customers think about their products and what kinds of events influence the company image. This goes far beyond the simple tracking of keywords and the monitoring tools of today.

💡 SOLUTION

BlogIntelligence

BlogIntelligence is a research prototype that tackles the problem of understanding social media. It supports the mining, analyzing, modeling, and presenting of blog data. With the underlying power of HANA, this tool is fast enough to interactively explore the blogosphere. Thus, users can find interesting information by focusing on specific questions and browsing interconnected blogs step by step. Here, a typical search for the term "SAP HANA" is used as an example.

1. Start with a standard search and find relations

As it is typical for search engines, the landing page of BlogIntelligence contains a search bar that highlights the user information needs. After entering a search string such as "SAP HANA," the user is immediately redirected to the overview page with related blog posts on the left and a visualization of related terms on the right (see Figure 11.6). This visualization can be used for further explorations of the dataset and makes it possible to see relations to other products, persons, and companies based on thousands of blog posts that concern the search term. The colors in the image reflect how often both terms occur together, from green representing frequent occurrence to orange as occasional. With these visualization capabilities, companies can understand what social media users relate to each search query and how the thematic search space has evolved to previously unknown associations.

2. Analyze trends

After the initial search, the BlogIntelligence trends view gives users a feeling for the evolution of discussions concerning their search terms (see

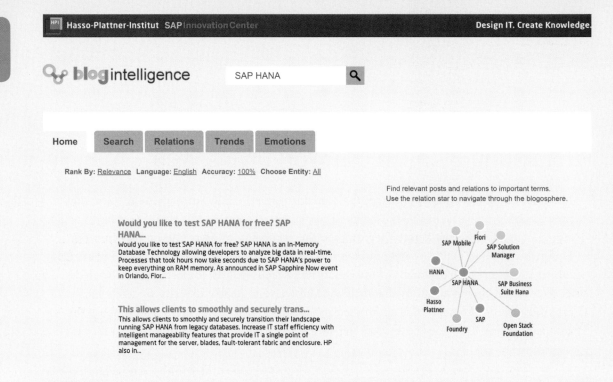

FIGURE 11.6
The BlogIntelligence Portal with the search term "SAP HANA"

Figure 11.7). In this example, we analyze the activities of the term around the SAPPHIRE '14 conference and look at the relationships between the three terms (HANA, SAP, and SAPPHIRE). The colors show the occurrences of the two specific search terms (red – HANA, yellow – SAP) and their related term (orange – SAPPHIRE). The peak on June 4, 2014 is noticeable and directly corresponds to the keynote events. Thus, we can find corresponding patterns to events which have influenced discussions about the original query. By incrementally adding more terms, users can further compare the evolution of different topics with their related search term. On top of this, BlogIntelligence can detect outliers and indicate whether peaks deviate from expected social media behavior.

3. Classify emotions about the topic

BlogIntelligence further offers a map of emotions. This feature helps users to understand blog data by integrating sentiments into the analysis. Users do not only see whether blogs include positive or negative statements, but also get an impression of when the most controversial or one-sided discussions are taking place.

With all these features, BlogIntelligence allows companies to better understand public opinion about their products. As a result, they can quickly identify patterns in customers' discussions that reveal new insights which help run their business more smoothly. Starting with a visualization that highlights sentiments per day, users can dig into each sentence that influences the public opinion.

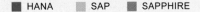

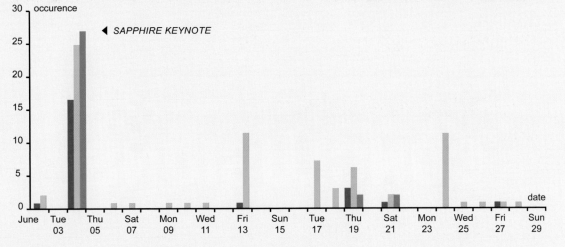

FIGURE 11.7
The trends view shows the evolution of related search terms. Here, the analysis is limited to German blog posts about SAP HANA around the SAPPHIRE '14 conference (June 3–5, 2014).

ENABLEMENT BY HANA

In-Memory Text Analysis

While the large amounts of blog data exceed the abilities of traditional databases, HANA is able to handle such loads very efficiently [HBM13]. For example, the flexible and interactive search requires complex aggregations of extracted blog metadata. As this is one of HANA's strengths, searching and ranking of terms is fast and still allows users to drill down to single sentences. To process the unstructured information of blogs, HANA offers integrated text analysis (see Section 2.2: Advanced Features). This feature makes it possible to execute text mining algorithms in memory, directly on the extracted blog data. Thus, BlogIntelligence is capable of processing millions of publicly available social media documents, containing several hundred million terms, within milliseconds. For a dataset of ten million blog posts, a traditional database takes up to two minutes to answer a search query – HANA is able to process the same search query within about 22 milliseconds. Furthermore, with its smart text mining and document analysis features, HANA can calculate and predict trends for every unknown topic in less than one second. This means that users can immediately see the results of their search query and gain fast insights from large amounts of unstructured data.

CHAPTER TWELVE
CONCLUSIONS

After riding successive waves of innovation with R/2, R/3, and ECC 6.0, SAP has once more changed the game in enterprise applications with the new Business Suite. S/4HANA combines a simplified data model, the Fiori User Experience (UX), and guided configuration. In the continuously evolving field of enterprise applications, these innovations can only be the beginning of a longer trajectory.

Let us take a look into some of the major challenges that lie ahead:

The Cloud

One of the dominating trends in the market for enterprise applications is the move towards the Cloud. The business model of S/4HANA reflects this trend (see Chapter 4: The New Business Suite S/4HANA). Its latest enterprise applications are first released to the Cloud, where new analytical, read-only applications appear without disruption. Because of SAP's non-disruptive strategy, companies that decide to keep their data on-premise can run new applications in parallel to their existing ones, assuming they already run on HANA.

Big Data

The amount of available enterprise data is growing rapidly and state-of-the-art database platforms such as HANA have to enable real-time Big Data processing. As computation continues to move closer to the data, extensive libraries of services in simulation, prediction, pattern recognition, time-series processing, and geospatial information processing become standard (see Chapter 11: Solving the Unsolvable). To guarantee the capacity to process Big Data, HANA balances the two approaches of scale-up and scale-out by making best use of replication and smart data partitioning according to the application logic (see Chapter 2: The Design Principles of an In-Memory Columnar Storage Database). The actual part of the data, which can still be transactionally modified, is kept in the main in-memory system following the scale-up approach. In contrast, the historical part of the data, which is read-only, can be scaled out to cloud clusters via replication. This form of data partitioning also lends itself to the exploitation of new storage technologies.

Real-time business

The volatility of markets remains high for the foreseeable future, and businesses have to deal with possible large swings in costs. Unanticipated consequences from technology shifts, such as the boom in US shale gas and oil, continue to shift energy fundamentals. This again underlines the critical need for flexibility. To stay at the head of their industries, companies need to minimize the time from observing signals in the market and in their organization to acting on these signals. The speed of HANA helps them in this endeavor to stay one step ahead of the competition (see Chapter 9: Exploiting the Window of Opportunity).

Products as a Service and the Internet of Things

The transition from selling tangible products to selling equivalent services has already begun, allowing companies to optimize the productivity of assets throughout their lifecycles and access new revenue opportunities. This trend is supported by the collaborative consumption (resource sharing) that is becoming common practice amongst consumers, especially in cities. Smart devices will connect companies to their customers (see Section 11.1: Smart Vending) and predictive maintenance will become a common business model to replace selling products by selling services (see Section 11.2: Remote Service and Predictive Maintenance).

Omni-channel commerce

Over the past decade, the ubiquitous accessibility of information and communication on mobile devices has significantly changed consumer

Complexity is the number one obstacle to innovation.

behavior. Overall, people now make more informed, conscious buying decisions. The key to facing this shift in consumer behavior is the ability to analyze the signals of societal change. This may be achieved by statistical customer segmenting (see Section 5.4: Customer Segmentation for HSE24 and Section 10.1: Determining the Who, What, and Where of Consumer Behavior) and investigating public opinion in social media (see Section 11.5: Understanding the Opinion of the Digital Society). After the different customer segments are identified and understood, they can be targeted on an individual level via the optimal sales and communication channel (see Section 9.1: Omni-Channel Retailing).

Business networks

Barriers to entry markets decline while business networks and startups proliferate. The advent of web, mobile, and social applications, as well as the explosive growth of business networks, is leveling the playing ground for smaller players. SAP has understood this trend and integrated the cloud services from Ariba (see Section 10.4: Driving Sourcing and Procurement Excellence through Better Spend Visibility), Fieldglass, and Concur into its existing business network.

Knowledge work

Computers are now becoming capable of doing jobs that it was once assumed only humans could perform. Machines can now act on unstructured commands and make subtle judgments. This results in both an extension of the power of human workers and in the offloading of tedious work. Knowledge work automation tools and systems will take on tasks that would be equivalent to the output of 110 million to 140 million full-time workers [McK13]. SAP is already taking the first steps towards this direction with automated fraud detection (see Section 8.3: Fraud Management) and risk analysis (see Section 9.3: Risk Analysis During Natural Disasters) based on the predictive algorithms embedded in HANA.

Processes optimization

Complexity is the number one obstacle to innovation. To tackle complexity, it is essential to remove unnecessary dependencies between different roles in the organization. The role-based enterprise applications built on HANA achieve this goal by empowering the users to access all information necessary to get their jobs done (see Chapter 10: Quality Time at Work). A second prerequisite for simplifying the organizational structure of an organization is to understand its processes (see Section 7.3: Celonis Process Mining – A New Level of Process Transparency). Once the inefficient organizational structure of a business process has been improved, the enterprise system has to be flexible enough to accommodate the changes (see Chapter 7: Reduction of Complexity and Chapter 8: Highest Flexibility by Finest Granularity).

With S/4HANA, SAP offers its customers a non-disruptive way to profit from the disruptive innovations of the HANA technology, be it in-cloud or on-premise. To address the challenges we discussed above, the HANA platform is built as a universal data-hub. For this, HANA natively embraces structured and unstructured data and integrates internal and external data.

HANA can integrate structured Enterprise Resource Planning (ERP) data with external data sources, such as:
> sensor data from smart devices in the Internet of Things
> data from social networks and the Internet
> data from the SAP Business Network

On top of the integration, HANA provides built-in algorithms to analyze this data on the highest level of granularity.

For years, enterprise systems were built to help companies run better. This was under the assumption that we, the enterprise system architects, could anticipate the needs of the respective industries and lines of business. We materialized these anticipations literally in predefined data aggregates. This was the right choice in a time of moderate change, when companies would rely on a solid organizational structure to guarantee quality and incremental product improvements to generate growth. Those times are over. In the constantly changing market environment of today, we can no longer anticipate every industry development of tomorrow. With the power of HANA, SAP enables companies to not only react to business developments, but also intelligently analyze market trends in order to drive forward their business.

Acronyms

ACID Atomicity, Consistency, Isolation, Durability

ADT Audience Discovery and Targeting

AP Accounts Payable

API Application Programming Interface

APO Advanced Planning and Optimization

AR Accounts Receivable

ASCO American Society of Clinical Oncology

ASE Adaptive Server Enterprise

ATP Available-to-Promise

B1 Business One

BFL Business Function Library

BI Business Intelligence

BPC Business Planning and Consolidation

BW Business Warehouse

BWA Business Warehouse Accelerator

BYD Business by Design

CAR Customer Activity Repository

CEI Customer Engagement Intelligence

CEO Chief Executive Officer

CFO Chief Financial Officer

CO Controlling

COGS Cost of Goods Sold

COO Chief Operating Officer

CPG Consumer Packaged Good

CRM Customer Relationship Management

D&B Dun & Bradstreet

DIMM Dual Inline Memory Module

DRAM Dynamic Random-Access Memory

DSO Days Sales Outstanding

EAV Entity Attribute Value

ECC 6.0 SAP ERP Central Component 6.0

EIM Enterprise Information Management

EPM Enterprise Performance Management

ERP Enterprise Resource Planning

ETL Extract-Transform-Load

EWM Extended Warehouse Management

FI Financial Accounting

FICO Financial Accounting & Controlling

FKOM Field Kick-Off Meeting

GAAP Generally Accepted Accounting Principles

GIS Geographic Information System

G/L General Ledger

GMV Gross Merchandise Value

GPS Global Positioning System

GRC Governance, Risk, and Compliance

HCI HANA Cloud Integration

HCM Human Capital Management

HCP HANA Cloud Platform

HDFS Hadoop Distributed File System

HEC HANA Enterprise Cloud

HPA Hamburg Port Authority

HPDS Hierarchical, Packaging-aware Discovery Service

HPI Hasso Plattner Institute

HR Human Resources

HSE24 Home Shopping Europe 24

IBP Integrated Business Planning

ICI Imperial Chemical Industries

IFRS International Financial Reporting Standards

IMS Information Management System

I/O Input/Output

IoT Internet of Things

KPI Key Performance Indicator

LDP Labor Demand Planning

LoB Line of Business

LRU Least Recently Used

M&A Mergers & Acquisitions

MDX Multidimensional Expressions

ML Material Ledger

MIT Massachusetts Institute of Technology

MMA Margin Management and Analytics

MRI Medical Research Insights

MRP Material Requirements Planning

NCT National Center for Tumor Diseases

NUMA Non-Uniform Memory Access

OECD Organization for Economic Co-operation and Development

OLAP Online Analytical Processing

OLTP Online Transaction Processing
PaaS Platform as a Service
PAL Predictive Analysis Library
P&L Profit & Loss
PLM Product Lifecycle Management
PoS Point of Sale
RDS Rapid Deployment Solution
REST Representational State Transfer
ROI Return on Investment
SaaS Software as a Service
SCI Social Contact Intelligence
SCIC Supply Chain Info Center
SCM Supply Chain Management
SDA Smart Data Access
SFA Sales Force Automation
SMP Symmetric Multiprocessing
S&OP Sales & Operations Planning
SRM Supplier Relationship Management
TCI Total Cost of Implementation
TCO Total Cost of Ownership
TDCA Total Delivered Cost Analytics
TMM Total Margin Management
UI User Interface
UPI Unique Product Identification
UX User Experience
VDM Virtual Data Model

Bibliography

[Ass14] Association of Certified Fraud Examiners, Inc. Report to the Nations on Occupational Fraud and Abuse. 2014.

[Bic97] Peter Bickford. Human Interface Online: Worth the Wait? Netscape DevEdge, 1997.

[CM01] Andy Cockburn and Bruce McKenzie. What Do Web Users Do? An Empirical Analysis of Web Use. International Journal of Human-Computer Studies, 54(6), 2001.

[Cod90] Edgar F. Codd. The Relational Model for Database Management: Version 2. Addison-Wesley Longman Publishing Co., Inc., 1990.

[CP13] Amy Chozick and Nicole Perlroth. Twitter Speaks, Markets Listen and Fears Rise. New York Times, 2013.

[Eva11] Dave Evans. The Internet of Things: How the Next Evolution of the Internet is Changing Everything. CISCO Systems, Inc., 2011. White Paper.

[FSKP12] Martin Faust, David Schwalb, Jens Krüger, and Hasso Plattner. Fast Lookups for In-Memory Column Stores: Group-Key Indices, Lookup and Maintenance. In ADMS (in conjunction with VLDB), 2012.

[GMR96] Bill Gates, Nathan Myhrvold, and Peter Rinearson. The Road Ahead. Penguin UK, 1996.

[GN00] Angelico A. Groppelli and Ehsan Nikbakht. Finance. Barron's Business Review Series. Barron's Educational Series, Inc., 4th edition, 2000.

[Ham14] Hamburg Port Authority. Port of Hamburg – Digital Gateway to the World, 2014. Brochure.

[HBM13] Patrick Hennig, Philipp Berger, and Christoph Meinel. Blog-Intelligence Extension with SAP HANA. In HPI Future SOC Lab: Proceedings 2013, 2013.

[IBM10] IBM Institute for Business Value. The Smarter Supply Chain of the Future, 2010. Report.

[KKG+11] Jens Krüger, Changkyu Kim, Martin Grund, Nadathur Satish, David Schwalb, Jatin Chhugani, Pradeep Dubey, Hasso Plattner, and Alexander Zeier. Fast Updates on Read-Optimized Databases Using Multi-Core CPUs. PVLDB, 5(1), 2011.

[KM58] Edward L. Kaplan and Paul Meier. Nonparametric Estimation from Incomplete Observations. Journal of the American Statistical Association, 53(282), 1958.

[Krü15] Jens Krüger. SAP Simple Finance: An Introduction. Galileo Press, Inc., 2015.

[MBBH15] Christoph Meinel, Justus Broß, Philipp Berger, and Patrick Hennig. Blogosphere and its Exploration. Springer, 2015.

[MCB+13] James Manyika, Michael Chui, Jacques Bughin, Richard Dobbs, Peter Bisson, and Alex Marrs. Disruptive technologies: Advances that will transform life, business, and the global economy. McKinsey Global Institute, 2013. Report.

[Mei14] Jens Meier. Smartening up to the occasion. portstrategy, 2014.

[Mil68] Robert B. Miller. Response Time in Man-computer Conversational Transactions. In AFIPS Fall Joint Computer Conference, volume 33. ACM, 1968.

[Mül13] Jürgen Müller. A Real-Time In-Memory Discovery Service: Leveraging Hierarchical Packaging Information in a Unique Identifier Network to Retrieve Track and Trace Information. Springer, 2013.

[Mul14] Joe Mullich. Using Real-Time Insights, HSE24 Gets Closer to Customers. Bloomberg Businessweek Research Services, 2014.

[OEC09] OECD. Magnitude of Counterfeiting and Piracy of Tangible Products: An Update, 2009. Report.

[Par14] Shaheen Parks. Projected Cost Analysis of the SAP HANA Platform: Cost Savings Enabled By Transitioning to the SAP HANA Platform, 2014. Report.

[Pla09] Hasso Plattner. A Common Database Approach for OLTP and OLAP Using an In-Memory Column Database. In Proceedings of the 35th SIGMOD International Conference on Management of Data. ACM, 2009.

[Pla13] Hasso Plattner. A Course in In-Memory Data Management. Springer, 2013.

[PMW09] Hasso Plattner, Christoph Meinel, and Ulrich Weinberg. Design Thinking – Innovation lernen – Ideenwelten öffnen. mi-Wirtschaftsbuch, 2009.

[SAP14a] SAP. Annual Report 2013. Web. March 2014.

[SAP14b] SAP. Real-Time Enterprise Stories. http://www.sap.com/bin/sapcom/en_us/downloadasset.2014-10-oct-13-21.sap-hana-real-time-enterprise-stories-pdf.bypassReg.html, 2014. Accessed on 2015-03-10.

[Sco12] Colin Scott. Latency Numbers Every Programmer Should Know. http://www.eecs.berkeley.edu/~rcs/research/interactive_latency.html, 2012. Accessed on 2015-03-10.

[Shn84] Ben Shneiderman. Response Time and Display Rate in Human Performance with Computers. ACM Computing Surveys, 16(3), 1984.

[SHS+15] Rita L. Sallam, Bill Hostmann, Kurt Schlegel, Joao Tapadinhas, Josh Parenteau, and Thomas W. Oestreich. Magic Quadrant for Business Intelligence and Analytics Platforms. Gartner, 2015. Report.

[Wal13] Bryan Walsh. A Year After Sandy, Living Dangerously by the Sea. Time Magazine, 2013.

Printing and Binding: Stürtz GmbH, Würzburg